T0235781

Clear Speech

Technologies that Enable the Expression and Reception of Language

Synthesis Lectures on Assistive, Rehabilitative, and Health-Preserving Technologies

Editor
Ronald M. Baecker, *University of Toronto*

Advances in medicine allow us to live longer, despite the assaults on our bodies from war, environmental damage, and natural disasters. The result is that many of us survive for years or decades with increasing difficulties in tasks such as seeing, hearing, moving, planning, remembering, and communicating.

This series provides current state-of-the-art overviews of key topics in the burgeoning field of assistive technologies. We take a broad view of this field, giving attention not only to prosthetics that compensate for impaired capabilities, but to methods for rehabilitating or restoring function, as well as protective interventions that enable individuals to be healthy for longer periods of time throughout the lifespan. Our emphasis is in the role of information and communications technologies in prosthetics, rehabilitation, and disease prevention.

Fieldwork for Healthcare: Case Studies Investigating Human Factors in Computing Systems
Dominic Furniss, Aisling Ann O'Kane, Rebecca Randell, Svetlena Taneva, Helena Mentis, and Ann Blandford
2014

Interactive Technologies for Autism
Julie A. Kientz, Matthew S. Goodwin, Gillian R. Hayes, and Gregory D. Abowd
2013

Patient-Centered Design of Cognitive Assistive Technology for Traumatic Brain Injury Telerehabilitation
Elliot Cole
2013

Zero Effort Technologies: Considerations, Challenges, and Use in Health, Wellness, and Rehabilitation
Alex Mihailidis, Jennifer Boger, Jesse Hoey, and Tizneem Jiancaro
2011

Design and the Digital Divide: Insights from 40 Years in Computer Support for Older and Disabled People
Alan F. Newell
2011

Clear Speech: Technologies that Enable the Expression and Reception of Language
Frank Rudzicz

ISBN: 978-3-031-00471-1   paperback
ISBN: 978-3-031-01599-1   ebook

DOI 10.1007/978-3-031-01599-1

A Publication in the Springer series
SYNTHESIS LECTURES ON ASSISTIVE, REHABILITATIVE, AND HEALTH-PRESERVING TECHNOLOGIES

Lecture #8
Series Editor: Ronald M. Baecker, *University of Toronto*
Series ISSN
Print 2162-7258   Electronic 2162-7266

Clear Speech

Technologies that Enable the Expression and

Reception of Language

Frank Rudzicz

Toronto Rehabilitation Institute and
Department of Computer Science, University of Toronto

*SYNTHESIS LECTURES ON ASSISTIVE, REHABILITATIVE, AND
HEALTH-PRESERVING TECHNOLOGIES #8*

ABSTRACT

Approximately 10% of North Americans have some communication disorder. These can be physical as in cerebral palsy and Parkinson's disease, cognitive as in Alzheimer's disease and dementia generally, or both physical and cognitive as in stroke. In fact, deteriorations in language are often the early hallmarks of broader diseases associated with older age, which is especially relevant since aging populations across many nations will result in a drastic increase in the prevalence of these types of disorders. A significant change to how healthcare is administered, brought on by these aging populations, will increase the workload of speech-language pathologists, therapists, and caregivers who are often already overloaded.

Fortunately, modern speech technology, such as automatic speech recognition, has matured to the point where it can now have a profound positive impact on the lives of millions of people living with various types of disorders. This book serves as a common ground for two communities: clinical linguists (e.g., speech-language pathologists) and technologists (e.g., computer scientists). This book examines the neurological and physical causes of several speech disorders and their clinical effects, and demonstrates how modern technology can be used in practice to manage those effects and improve one's quality of life. This book is intended for a broad audience, from undergraduates to more senior researchers, as well as to users of these technologies and their therapists.

KEYWORDS

computational linguistics, speech-language pathology, assistive technologies, rehabilitation science, machine learning

Dedicated to those overcoming barriers in communication

Contents

Preface

When I was a grad student, the purpose of my research was to improve the accuracy of speech recognition software for people with speech disorders. I started by working with cerebral palsy (CP), which remains the most common cause of hard-to-understand speech today. Most people with CP were not very well understood by speech recognition at the time—less than 1% of their words could be correctly recognized whereas a speaker without a speech disorder might be comfortably understood 85% or 90% of the time. It wasn't that their words didn't make sense—people with CP can normally understand and produce *language* just fine—it was that their voices are quite different from those of the general population, which can profoundly confuse speech software. It was my job to un-confuse the software.

Not being understood *almost all of the time* can be annoying in itself—and speech recognition certainly did a dismal job for people with CP. It was therefore perhaps somewhat frustrating that speech was often the *most effective* means of communication these individuals had. Although CP limits the control of the muscles of speaking (e.g., the tongue), CP *also* affects other muscles (e.g., those controlling the fingers). This means that while speech in CP can be approximately three times slower than typical speech, typing can be over a *hundred* times slower.

So if a computer can't understand what you say *and* it takes too long or is too difficult to type by hand, then merely participating in our modern society becomes a tremendous challenge. According to the U.S. Census bureau, less than 10% of people with severe disabilities are employed, partially due to difficulty in communication, which has considerable consequences for social and health well-being.

Something must be done.

So how could I make my own small dent toward cracking this huge problem? Since the *sounds* of speech in cerebral palsy were so difficult for computers to understand, I reasoned that it might help to "teach" the computer *why* those sounds were difficult—to teach it about differences in the *physical* origins of speech. How do you teach a computer? These days, we use MACHINE LEARNING where you basically program the computer to find patterns and relationships in data by itself, typically given lots of carefully curated examples that you provide. In my case, I needed to provide examples of speech sounds and their corresponding vocal tract movement, and for that I needed participants to come into the lab to have their voices and facial movements recorded during speech.

Many of the participants were in their early twenties and came in with their parents or other caregivers. One young man with CP was particularly talkative, and his father was equally eager to insert himself into the conversation, usually to repeat or to clarify what his son said. They were both very outgoing, and we had about as non-serious a chat as you can imagine in a research setting, in the basement of a satellite building of the University of Toronto. At one point, the young man revealed that one of his main motivations for volunteering (and for getting his dad to take time off of work to drive him into the lab), was "girls." I told him that was not part of our research protocol. This young man's father then chimed in to say that it wasn't so much "girls" as it was a *particular* girl, and that she and his son were "courting,"[1] but communication between them remained difficult. The young man had tried a number of devices and programs to help him be understood, but he found each of them to be insufficient—he didn't feel like he could properly express himself. The alternative to talking through a computer was to talk through the filter of your father, which can also be non-ideal in courtship. He wanted to help us improve the technology.

Can advanced speech technology improve your love life? More data is required. However, what was clear to me from that exchange was that so much of who we are, collectively and as individuals, depends on our ability to communicate. Language is not just about communicating facts or making plans—to a large extent it defines how others perceive us and how we perceive ourselves in the world. Being able to define yourself in your own words—to speak for yourself—is liberating.

I hope that this book can bring together people who really should be talking together, especially technologists, therapists and clinicians, and people affected by speech disorders. Technologists need to know what challenges exist in the real world and how clinicians are currently meeting those challenges. Therapists need to know how artificial intelligence that can help to diagnose, monitor, and overcome issues of communication. Perhaps most importantly, people affected by speech and language disorders need to know that there is light at the end of the tunnel, and that technology is helping to provide that light.

> [Language has a] unique role in capturing the breadth of human thought and endeavour...We look back at the thoughts of our predecessors, and find we can see only as far as language lets us see. We look forward in time, and find we can plan only through language. We look outward in space, and send symbols of communication along with our spacecraft, to explain who we are, in case there is anyone there who wants to know. [Crystal, 1998]

Frank Rudzicz
February 2016

---

[1] Is that what kids do these days?

Figure Credits

CHAPTER 1

Introduction

About one in every ten people in the world, from newborns to the oldest among us, has some communication disorder affecting speech. These disorders can manifest themselves physically (as in reduced control of the muscles of speech in cerebral palsy and Parkinson's disease), cognitively (as in difficulty understanding words in autism and dyslexia), or both physically and cognitively (following, for example, cardiovascular stroke), according to the U.S. National Institutes of Health. These figures are increasing with the age of the population and the incidence of stroke. The Canadian Association of Speech-Language Pathologists and Audiologists estimates that one in ten people are affected to some degree by language impairments and that this proportion will rise significantly over the next decade with the prevalence of cardiovascular stroke expected to rise as populations in various countries become older. In fact, speech and language disorders are present in nearly 85% of those who have experienced stroke and are one of the first symptoms of Alzheimer's disease. This prevalence is especially worrying since aging populations across many nations will result in a drastic increase in speech disorders brought on by age. This will place a tremendous burden on speech-language pathologists, therapists, and caregivers who are often *already* overworked or, in many cases, devoted to language disorders that occur *earlier* in life, such as in cerebral palsy or in developmental delays. This has been referred to as an impending healthcare crisis. At the very least, it will require massive changes in how healthcare is delivered, globally.

Fortunately, modern technology has matured to the point where it can now have a profound positive impact on the lives of millions of people with speech and language disorders. This book serves as common ground for several communities, especially clinical linguists (including speech-language pathologists), and technologists (including computer scientists and engineers). Hopefully, sharing common ground will help to accelerate collaboration in this area. The book is, however, written for a broad audience, from advanced undergraduates and more senior researchers to users of assistive technologies, their families, and their therapists.

Before we continue, we should make a few terminological clarifications. To be properly pedantic, we should distinguish between SPOKEN LANGUAGE (referring to word-, grammar-, and meaning-level aspects of language in spoken utterances), WRITTEN LANGUAGE (referring to those aspects in writing), and SPEECH (referring to acoustics and articulatory aspects of speech acts). While it is important to be cognizant of the differences between these terms, we will occasionally use the term "speech" as a superset of its physical and acoustic properties *and* the linguistic aspects of speech acts. We should also be clear as to the scope of this book. We will not cover all topics in speech and language, naturally. Our focus is on technologies that assist in speech communication.

This will include technologies that can interpret difficult speech and those that can synthesize easy-to-understand speech. Along these lines, we will also discuss the entry of written text to drive those systems.

Part I provides some mathematical and terminological background to help interpret the rest of the book. Not all of Part I will be applicable to you, but if you are a technologist missing a background in linguistics, or a linguist without experience in modern machine learning, these chapters will at least help you to communicate in the same language (so to speak) as the other people in the room. Part II covers the NEUROLOGY AND ANATOMY of speech and language; to a large extent, this covers information about how the brain processes and produces language, and what can go wrong in the vocal tract and hearing mechanism. While the focus will be on speech, we will also take a brief foray into cognitive disorders affecting language comprehension.

Part III covers TECHNOLOGIES THAT ENABLE, especially those that speak for people with impairments of speech production (including eye-typing and word prediction), and those that help interpret for those with impairments of speech reception (including cochlear implants and other hearing aids). This will be a whirlwind tour of this area of research and will in many ways only scratch the surface. You are therefore invited to follow the various citations and references that will be provided throughout these sections.

PART I

Background

CHAPTER 2

Math & Stats for Language Technology

Like so much of human behavior, language opens itself up to formal analysis by mathematics and statistical processes. This includes everything from describing the motions of the physical articulators to mimicking auditory processes in artificial neurons. In order to make meaningful progress in our field of research, we often require a thorough familiarity with various statistical tools. This chapter surveys mathematical approaches that are relevant to certain subareas within speech and language processing in a fairly introductory manner, using examples relating to our domain. By no means is this survey exhaustive—for almost every tool we discuss, you will find more intricate varieties that can be more suitable to your task, so you are encouraged to dig deeper by following the included references and by doing your own research.

We begin by discussing basic probability theory, which is a central component in modern computer systems of language, which use statistics to make interpretations. Probability theory is also a central component of INFORMATION THEORY, which concerns the uncertainties present in the transmission of information between abstract producers and receivers of messages. Resolving and modeling these uncertainties using statistical probabilies can be an important component of speech technologies.

2.1 PROBABILITY THEORY

PROBABILITY THEORY deals with representing the likelihood of events. The canonical examples involve games—how likely are you to roll a 4 with a fair 6-sided die? How likely are you to pull the Queen of Hearts from a deck of playing cards? In these examples, we're using probability theory for one of its chief purposes—to assign a likelihood to a particular EVENT. In the fair die example, the event is that a particular side, $A = 4$ will turn up, but there are several *possible* events (in this case, six). We often call the list of all possible events the SAMPLE SPACE or DOMAIN, represented as Ω, where the number of elements in that set are denoted by double bars. For example, if Ω is sides on a die, $\|\Omega\| = 6$.

We can often think of language in similar terms. Imagine that our domain Ω is the set of all English words, of which there are approximately $\|\Omega\| = 250,000$.[1] You may have stumbled on it or stepped over it, but there was an AMBIGUITY in the previous sentence, which is an important

---

[1]According to the Oxford English Dictionary, which generally includes only common, non-slang words. Technical terms, including medical/clinical language, greatly expand this set.

concept in computational linguistics.[2] The ambiguity is around the meaning of the word "*word*". Sometimes, the word "*word*" is synonymous with TERM, which is like an entry in a dictionary of which there are, as estimated, about 250,000 in English. Other times, the word "*word*" means a sequence of characters separated by spaces—an instance of a term—of which there have been countless trillions scribbled down over the centuries. An *instance* of a word is called a TOKEN. For example,

Counting terms $Since_1$ the_2 $terms_3$ "the", "in_4", "$this_5$", "terms", and_6 "$sentence_7$" are_8 $repeated_9$ in this sentence, the $number_{10}$ of_{11} *terms* in this sentence is_{12} $thirteen_{13}$.

Counting tokens $This_1$ $sentence_2$ by_3 $contrast_4$, has_5 $seven_6$ $tokens_7$, no_8 $wait_9$, $make_{10}$ $that_{11}$ $thirteen_{12}$ $tokens_{13}$.

We'll return to ambiguity in Chapter 3, so this digression is merely to emphasize that when dealing with probabilities, it's important to carefully define your domain first.

Given a domain, we now have the task of defining how likely a given event is. In the case of a fair 6-sided die, the probability of any one side turning up should be the same for every side, i.e.,

$$P(A = 1) = P(A = 2) = \ldots = P(A = 6) = P(A) = 1/6.$$

To a large extent, research in natural language by computers treats language itself as no more than an $\|\Omega\|$-sided die, which we metaphorically roll each time we write a word on a page (or, more likely these days, a keyboard). Imagine you have a very limited vocabulary of 25 terms, $\Omega = \{the, of, in, house, cat, hat, \ldots, pulchritudinous\},$[3] where each word is equally likely. This is what we call a "flat" or UNIFORM DISTRIBUTION, so called because the resulting probability distribution of the words, shown in Figure 2.1, is uniformly flat.

Now, naturally not all words in a human language are equally likely. You are far more likely to write the word "the" than you are to write the word "pulchritudinous". Here, the metaphor of language-as-dice must be amended, but it can still work. *Loaded* dice are dice that are weighted assymetrically so that certain sides are more likely to come out on top than others. If you are handed a loaded die and tasked with discovering its BIAS, that is, the probabilities of each side, a good approach would be to roll the die a very large number of times, count the occurrances of each side, and divide each of those counts by the total number of rolls to get the probabilities of each side. The same can be done with language. Using the loaded die metaphor, we can imagine that every word token we see on a page (or on a screen, or uttered by a mouth, etc.) is the result of a "roll of the language die". If an imaginary book has a million word tokens ($N = 1 \times 10^6$) and 40,000 of those are the term "the" (i.e., we *Count(the)* forty thousand times), then the PRIOR

---

[2]Computational linguistics is a field of research that processes natural language using algorithms and statistics, and is discussed in Chapter 3.

[3]**pulchritudinous** *adj.* Beautiful. **Origin** 1910–15 Americanism.

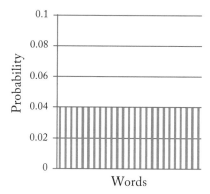

Figure 2.1: A uniform distribution where each bar represents the probability of each of 25 equally likely words, i.e., each word in this limited vocabulary has a 4% probability of being said.

PROBABILITY of a word W being "the" is

$$P(W = the) = Count\,(the)/N = \frac{40,000}{1 \times 10^6} \approx 0.04.$$

As in all probabilistic models, we are bound by the rules of probability. The two chief rules any probabilistic model must obey are:

- The probability of any event A must be between 0.0 and 1.0 inclusive, i.e., $0.0 \le P(A) \le 1.0$. If $P(A) = 0$, A is impossible; if $P(A) = 1$, no event *other* than A is possible.

- The sum of probabilities over all events (possible or otherwise) must sum to 1.0, i.e., $\sum_i P(A_i) = 1$.

Since every word token in our imaginary book must count toward exactly one of the terms used in that book, clearly $\sum_W P(W) = 1$.

One can estimate the probability of each word in a collection of text by simply counting the occurrence of each term and dividing by the total number of tokens in that collection. This constitutes a LANGUAGE MODEL, which is a centrally important concept in computational linguistics. At its most basic, it is a representation of the likelihood of words, which permits all manner of applications to be possible, from making more informed guesses in interpreting hard-to-understand speech to predicting the next word someone is trying to type. Representing the probabilities of words helps the computer to make smarter interpretations of language.

Like all models in science, a language model is a set of PARAMETERS that **describes** data that we've seen already and can **predict** future or unseen data. In Ptolemy's geocentric model of the solar system, the parameters were the widths of mystical transparent spheres on which the sun and planets revolved, convolutedly, around the Earth. In our language model, the parameters are merely the probabilies of each word. Ptolemy's geocentric model could also be used to make predictions, such as the location of Mars in the sky on a particular future evening. Similarly, our simple language model can be used to make predictions of "future words", as we'll discuss in Section 3.1. For example, you could find the probabilities of all of the words in George R.R. Martin's series of novels, *A Song of Ice & Fire*, to predict each subsequent word in the (as yet unpublished) rest of the series. Whether that prediction is *accurate* is another matter altogether, to which we will similarly return in a future section.[4]

2.1.1 MULTIPLE EVENTS

Just as Ptolemy's Earth-centered model of our solar system turned out to be inadequate, so too will our simplistic model of language. Merely finding the probability of a word will encapsulate (almost) nothing with regard to grammar, meaning, or context. It's not possible to learn a language model from the first $N - 1$ chapters of a murder mystery in this way, for example, to accurately predict the culprit in the Nth chapter.

An account of modern computational theories of syntax, semantics, and pragmatics is beyond the scope of this book. However, for now it suffices to say that, even at our somewhat naïve level of language modeling, we can learn models that are slightly more complex by using more than just one RANDOM VARIABLE. There are many ways we can do this—we are really only limited by our imagination.[5] For example, for any word W, we could model its joint probability with the speaker being in a particular emotional state, E; $P(W = dang, E = angry)$ could be the JOINT PROBABILITY of uttering the word "dang" while simultaneously being angry. We're also not bound to using only two random variables, so long as we obey the laws of probability. Specifically, to generalize our earlier rules:

- The probability of any n joint events $x_1 \ldots x_n$ must be between 0.0 and 1.0 inclusive, i.e., $0.0 \leq P(x_1, \ldots, x_n) \leq 1.0$.

- The sum of probabilities over all events (possible or otherwise) must sum to 1.0, i.e., $\sum_{x_1} \sum_{x_2} \cdots \sum_{x_n} P(x_1, \ldots, x_n) = 1$.

Often, we use this notation not to refer to n events happening *simultaneously* but instead as a sequence of n single events occuring in strict succession. If x_1 is the first word in a sequence, x_2 is the second. Also, with multiple events, we can also introduce new rules. Perhaps the most familiar of these is the CHAIN RULE which is:

$$P(A, B) = P(B \mid A) P(A) \tag{2.1}$$

or, more generally,

$$P(x_1, x_2, ..., x_n) = P(x_1) P(x_2 \mid x_1) P(x_3 \mid x_1, x_2), \ldots, P(x_n \mid x_1, x_2, \ldots, x_{n-1}). \tag{2.2}$$

---

[4]Can you predict which one?
[5]And data!

When combined with the notion of a joint probability as a probability of a sequence, we can talk about things like:

$$
\begin{aligned}
P(w_1, w_2, w_3, w_4) &= P(a, long, time, ago) \\
&= P(w_1)P(w_2 \mid w_1)P(w_3 \mid w_1, w_2)P(w_4 \mid w_1, w_2, w_3) \qquad (2.3) \\
&= P(a)P(long \mid a)P(time \mid a, long)P(ago \mid a, long, time)
\end{aligned}
$$

where we describe the probability of reading the sequence *a long time ago* as the probability of seeing "a", times the probability of seeing "long" given that we just read the word "a", times the probability of seeing "time" given that we just read *a long*, and so on. Importantly, you are not bound to read the words left-to-right. Any permutation of reading order is permitted—the probability of a three-word sequence can be the prior probability of the second word times the probability of the first given the second, times the probability of the third given the first two. This allows us to use another new rule of joint probabilities, namely BAYES' RULE which states

$$
P(A \mid B) = \frac{P(B \mid A)P(A)}{P(B)}. \qquad (2.4)
$$

This relationship[6] is visualized in 2.2 and is not merely provided here for your amusement—Bayes' rule is fundamental to many kinds of modern machine learning algorithms.

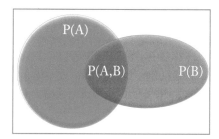

Figure 2.2: Bayes' rule represented in a standard Venn diagram of probabilities over random variables *A* and *B*.

2.2   INFORMATION THEORY

At some level, human communication is really about transmitting information, whether explicit words or a spectrum of intentions and emotions. Therefore, it can be useful to draw on mathematical theories collectively called INFORMATION THEORY to quantify the *amount* and type of information in a signal. Information theory dates to the end of the second World War and was initially a means to determine how to build in error-correction and redundancy given imperfect communication channels that could corrupt or distort your message [Shannon, 1949].

---

[6]You can work it out for yourself, using the chain rule, knowing that $P(A, B) = P(A)P(B|A) = P(B)P(A|B)$.

Let's go back to the metaphor of language as a die whose sides are words. Imagine we have an individual who can only say two words: *yes* and *no*—they are about to speak, and you're unsure as to which word they are about to utter. You have a certain *amount* of UNCERTAINTY—a lack of information. Imagine that this entity now utters the word *no*. Your uncertainty is gone and you've received information. How much? If *yes* and *no* are equally likely, then $P(no) = 0.5$ and therefore

$$I(E = no) = \log_2 \frac{1}{P(E = no)}$$
$$= \log_2 \frac{1}{\frac{1}{2}} = 1 \text{ bit.}$$
(2.5)

This might be intuitive—it takes a "bit", in computer terms, to encode a binary value. Not so surprising, after all. What if all this entity did was roll 6-sided dice instead? If the die is fair, each side is equally likely. How much information would we receive if the die came up with a 5 (whose probability is $P(E = 5) = 1/6$)?

$$I(E = 5) = \log_2 \frac{1}{P(E = 5)}$$
$$= \log_2 \frac{1}{\frac{1}{6}} \approx 2.59 \text{ bits.}$$
(2.6)

We receive more information when we observe the roll of a 6-sided die because there are more options—each possible outcome is less likely so we're more "surprised" when we observe a particular value in a sequence. Note that the base of the logarithm, 2 here, is not connected to the number of possible values. We could have chosen any base (e.g., base-10 or the natural base, e)—this value merely determines the units of information we're dealing with. Base-2 gives "bits" of information.

As we've previously discussed, the words in natural languages are not all equally likely. If you're wondering what the next word I might utter might be, you'll be far less surprised if the word is *the* than if it was *octogenarian*.[7] Fortunately, the same formulation above still holds. Even with a vocabulary of 150,000 words, if $P(E = the) = 0.05$, then $I(E = the) = \frac{1}{P(E=the)} = 20$ bits.

In general, the amount of information provided by a single symbol (e.g., word) is not as informative as the *average* amount of information provided by symbols observed from a system or entity over time, which leads us to the concept of ENTROPY. Before we discuss it formally in Section 2.2.1, let's find some entropy. Beyond representing an average amount of information gained, per symbol, for a system, entropy can in many ways characterize a distribution. Figure 2.3 shows two probability distributions over a set of words—one relatively flat and the other relatively "peaked". In many aspects of computational modeling of language, we prefer (in some sense) low entropy distributions because we can be more confident in our predictions—we have less uncertainty about what we will observe next in the sequence, on average.

---

[7]**octogenarian** *n.* someone who is between 80 and 89 years of age.

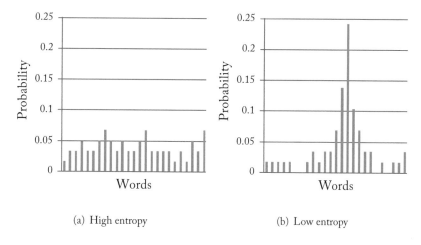

(a) High entropy   (b) Low entropy

Figure 2.3: Distributions with high (flatter) and low ("peakier") entropy.

Entropy is therefore equivalently:

1. The average amount of information provided by symbols in a vocabulary,

2. The average amount of uncertainty you have before observing a symbol from a vocabulary,

3. The average amount of "surprise" you receive when observing a symbol, and

4. The number of bits needed to communicate that alphabet.

2.2.1 ENTROPY

In many aspects of atypical speech, we will have considerably more statistical uncertainty than in typical speech. As discussed in Chapter 8, this is often the case in motor disorders such as cerebral palsy, where the control of the articulators might be less precise. We may wish to measure and compare the degree of statistical uncertainty in both acoustic and articulatory data for speakers with and without these disorders, as well as the *a posteriori* uncertainty of one type of data given the other. This quantification will inform us as to the relative merits of incorporating knowledge of articulatory behavior into speech technology systems for individuals with these disorders.

Entropy, $H(X)$, is a measure of the degree of uncertainty in a random variable X. When X is discrete, this value is computed with

$$H(X) = -\sum_{i=1}^{n} p(x_i) \log_b p(x_i),$$

where b is the logarithm base, x_i is a value of X, of which there are n possible, and $p(x_i)$ is its probability. When our observations are continuous, as they are in many acoustic and articulatory data, we must use *differential entropy* defined by

$$H(X) = -\int_X f(X) \log f(X) dX,$$

where $f(X)$ is the probability density function of X. For a number of distributions $f(X)$, the differential entropy has known forms [Lazo and Rathie, 1978]. For example, if $f(X)$ is a multivariate normal,

$$f_X(x_1, \ldots, x_N) = \frac{\exp\left(-\frac{1}{2}(x-\mu)^T \Sigma^{-1}(x-\mu)\right)}{(2\pi)^{N/2} |\Sigma|^{1/2}}$$

$$H(X) = \frac{1}{2} \ln\left((2\pi e)^N |\Sigma|\right), \tag{2.7}$$

where μ and Σ are the mean and covariances of the data. However, since we observe that both acoustic and articulatory data follow non-Gaussian distributions, we choose to represent these spaces by mixtures of Gaussians. Huber et al. [2008] have developed an accurate algorithm for estimating differential entropy of Gaussian mixtures based on iteratively merging Gaussians and the approximate upper bound of the entropy,

$$\tilde{H}(X) = \sum_{i=1}^{L} \omega_i \left(-\log \omega_i + \frac{1}{2} \log\left((2\pi e)^N |\Sigma_i|\right)\right),$$

where ω_i is the weight of the ith $(1 \le i \le L)$ Gaussian and Σ_i is that Gaussian's covariance matrix. Note that while differential entropies *can* be negative and not invariant under change of variables, other properties of entropy are retained [Huber et al., 2008], such as the chain rule for conditional entropy

$$H(Y \mid X) = H(Y, X) - H(X),$$

which describes the uncertainty in Y given knowledge of X, and the chain rule for mutual information

$$I(Y; X) = H(X) + H(Y) - H(X, Y),$$

which describes the mutual dependence between X and Y. Here, we quantize entropy with the *nat*, which is the natural logarithmic unit, e (≈ 1.44 bits).

These representations are very general, if also a bit technical, and can be useful in a wide variety of contexts. Representing language with this and similar information-theoretical models allows for a number of uses, including explaining how messages can be distorted as they are passed through a medium, such as speech over a telephone wire or Skype. This use, called the NOISY-CHANNEL MODEL, will be discussed in Section 8.1.7, in which we suggest that certain speech disorders can be explained by probabilistic distortions to the control signals that are passed from the brain to the articulators of speech (e.g., the tongue).

CHAPTER 3

(Computational) Linguistics

The field of COMPUTATIONAL LINGUISTICS (CL),[1] in broad terms, concerns getting computers to process human language. This pursuit has taken many forms, each of which has involved different challenges. The goal of creating a CONVERSATIONAL AGENT, in modern times, dates back at least to 1950 when Alan Turing proposed that the best way to determine if a machine *actually thinks* is to have a conversation with it [Turing, 1950] (through textual tele-type, the equivalent of modern text messaging, but the same principle applies).

Computational linguistics, by definition, must naturally be at the center of any system that processes language computationally, which includes tools to help individuals with linguistic disorders. The following subsections (very briefly) introduce core aspects of computational linguistics that are especially relevant in this application area. The first involves predicting words using statistics; while this is often the *goal* of many applications, this simple idea forms the foundation for many other aspects of language processing, including translating texts between languages and speech recognition. The second core component concerns linguistic *features*, which are numerical or quantitative measurements of specific aspects of a piece of text or speech. Accurate measurement of *relevant* features is often essential to the function of the third component, which is machine learning—a topic which extends beyond computational linguistics.

3.1 WORD PREDICTION

WORD PREDICTION is now a technology that almost anyone with a mobile phone has used—it is the automatic presentation of a list of possible continuations given what you've already typed, so you can save a bit of typing. Quite often (though not always!) modern phones will present the correct next word to you given only a few keystrokes. While the ubiquity of this technology may make it appear generic, it has tremendous implications for various populations with communication disorders. This assistive technology has reduced the number of keystrokes required of an individual by as much as $\sim 69\%$ in adaptive-lexicon systems [Matiasek et al., 2002, Swiffin et al., 1987], thereby increasing communication speed and allowing improved individual expression [Alm et al., 1992]. Prediction is especially valuable to those for whom fatigue or frustration often accompany attempts at communication [Garay-Vitoria and Abascal, 2006].

---

[1]The term "computational linguistics" is often synonymous with NATURAL LANGUAGE PROCESSING or HUMAN LANGUAGE TECH-NOLOGY, although there are partisans of each camp for whom the distinction, which generally involves the adherence to classical linguistics theory in CL, is paramount.

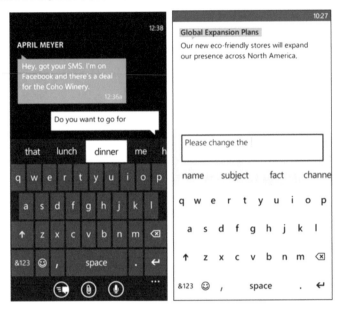

Figure 3.1: Word prediction on a modern mobile touch-screen device. Used with permission from Microsoft © 2016.

Word prediction can be accomplished easily by simply knowing the probabilities of sequences of word types. A sequence of N words, $w_1 \ldots w_N$, is called an N-GRAM and the probability of that N-gram is $P(w_1 \ldots w_N)$; often, we only have enough confidence in our statistics to look at small values of N, so we use the statistics of unigrams (1-grams, like *in*), bigrams (2-grams, like *in the*), or trigrams (3-grams, like *in the blue*). We can also transform N-gram probabilities to conditional probabilities.

If a user of an assistive communication device (or, indeed, any modern smartphone) has just typed *see you*, then we can use the trigram model to consider all possible words that come next. More formally, if $P(see\ you\ \mathbf{w}^*) \geq P(see\ you\ \mathbf{w_x})$ for any word $\mathbf{w_x}$, then $P(\mathbf{w}^* \mid see\ you) \geq P(\mathbf{w_x} \mid see\ you)$ and $\mathbf{w}^*$ is therefore the best next possible word, and we can suggest it to the user.

The probabilities of N-grams can be determined very easily. All that is required is a large sample of text data similar in nature to the kind of text we expect people to type. For example, if we expect people to type "financial news", we might learn our probabilities from the *Wall Street Journal*. If we expect people to type "medical diagnoses", we might learn our probabilities from medical textbooks. In either case, the data we gather is called a *corpus*,[2] sometimes denoted as C. Our probabilities can then simply be obtained by counting the occurrences of a particular N-gram in that corpus, divided by the total number of N-grams in that corpus. For example, if $N = 1$ and we want to know the prior probability of the word *octopus*, we would just count the number

---

[2]**corpora** *n.pl.* the plural of *corpus*.

of times *octopus* occurs in our corpus and divide it by the total number of 1-grams (i.e., words); if *octopus* occurs once in a corpus with 100 word tokens in total, we estimate $P(octopus) \approx 1/100 = 0.01 = 1\%$.

In general, this extremely simple approach works incredibly well and across contexts much more varied than word prediction. Knowing the probabilities of N-grams allows us to estimate the probabilities of much longer sequences. For example, if we have a bigram model, we can estimate the probability of longer sequences like *the cat in the hat* by multiplying together all the component bigram probabilities. That is:

$$P(the\ cat\ in\ the\ hat) \approx P(the\ cat) \cdot P(cat\ in) \cdot P(in\ the) \cdot P(the\ hat).$$

What this allows us to do is evaluate whether one sentence is more likely than another. In speech recognition, this means that if the system is having difficulty deciding between two competing hypotheses for what was said, we can rely on which hypothesis is simply more "likely" in a language. While this approach does have broad applicability, there are complications. Not least among these is the problem that everyday people utter or encounter sentences or even component phrases and N-grams that have never been uttered before. For example, you've probably never read or heard the sequence *Just Google the Instagram app*, despite its innocuousness. If you have never encountered the trigram *Google the Instagram* in your life (which constitutes a *very* large corpus of language), then our approach would estimate that $P(Google\ the\ Instagram) = 0$ and therefore that our nice little sample phrase is *impossible*. Fortunately, there are many algorithmic solutions. A gentle introduction to these can be found in the textbook by Jurafsky and Martin [2009].

The current word w_i can also be anticipated given an n-gram context augmented by part-of-speech tags t_j. For example, Fazly and Hirst [2003] describe an algorithm that ranks possible completions based on the estimate

$$
\begin{aligned}
P(w_i | w_{i-1}, t_{i-1}, t_{i-2}) &\approx \sum_{t_i \in T(w_i)} P(w_i | w_{i-1}, t_i) P(t_i | t_{i-1}, t_{i-2}) \\
&\approx \sum_{t_i \in T(w_i)} \frac{P(w_i | w_{i-1}) P(t_i | w_i)}{P(t_i)} P(t_i | t_{i-1}, t_{i-2}) \\
&= P(w_i | w_{i-1}) \sum_{t_i \in T(w_i)} \frac{P(t_i | t_{i-1}, t_{i-2}) P(t_i | w_i)}{P(t_i)}
\end{aligned}
\tag{3.1}
$$

where $T(w_i)$ is the set of all possible PoS tags associated with word w_i. Combining PoS with lexical context in this way reduces the percentage of keystrokes needed to produce text by $\sim 6\%$ over purely *a priori* statistical methods [Fazly and Hirst, 2003]. Other extensions to text prediction to further refine the list of hypothesized completions include the use of grammatical syntax and semantics [Erdogan et al., 2005, Li and Hirst, 2005], as well as trained neural networks [Garay-Vitoria and Abascal, 2006].

Empirically observed improvements in the rate of typed communication with prediction might not overcome improvements gained through the use of speech (see above), but applying the same approach to predicting spoken communication may reduce the amount of effort required for both the dysarthric speaker and their audience. If speech input is coupled with a visual display for output, for example, that display could be updated "on-the-fly" with the results of predicted queries before those queries are completed.

CHAPTER 4

Automatic Speech Recognition (ASR)

AUTOMATIC SPEECH RECOGNITION (ASR) is software that listens to speech signals and provides one or more textual hypotheses as to what was said. The popularity of this technology has risen and fallen since the 1960s, generally, but is currently robust enough to be used prominently by both desktop and mobile operating systems, at least for typical speakers. However, many challenges remain for *atypical* speakers. If the statistical models inherent to modern ASR are not built to accomodate speech differences, then speakers with those differences cannot expect to have their speech understood to the same extent as the general population. This can be a hindrance beyond merely not being able to ask Siri or Cortana for the weather. Indeed, individuals without the control of their arms can *rely* on speech recognition for *all* access to their computers, from transcribing emails to basic interaction with the system.

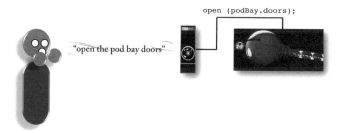

Figure 4.1: A successful communication of a directive statement by a human to a fictional spaceship's computer, using automatic speech recognition.

The goal of ASR is to decide on the optimal word sequence $W = w_1 \, w_2 \, ... \, w_n$ to describe an acoustic input speech signal X:

$$W_c = \underset{W}{\mathrm{argmax}} \, \frac{P(W)P(X|W)}{P(X)} \tag{4.1}$$

where $P(W)$ and $P(X|W)$ are the optimal language and acoustic models, respectively. The input speech signal, X, is typically measured at a constant sampling rate, where the ith measurement, $x[i]$, is quantized according to a constant bit rate (e.g., 8-bit values range from -128 to 127, 16-bit values from $-32,768$ to $32,767$). An example is shown in Figure 4.2. This discretized signal is first converted to an alternative form more amenable to machine learning via feature extraction and it is upon the space defined by these features that statistical models are trained and used in classifying sounds, as summarized below.

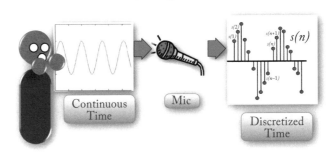

Figure 4.2: An abstract representation of the discretization of a speech signal in which the continuous changes in air pressure caused by speech over time are converted into a sequence of amplitude measures at discrete points in time.

Because certain words, W, are too infrequent to be learned directly from data, or because the number of *possible* words is too high (often over 60,000 in typical speech dictation tasks), ASR is usually built at the level of smaller chunks of speech called PHONEMES. A phoneme is perceptually distinct units of sound in a language; for example, the word *head* is composed of three phoneme instances, /h ∈ d/ and *heed* is also composed of three, /h i: d/.[1] Typically, an ASR system will have one model for each phoneme, will interpret speech as a sequence of phonemes,[2] and then partition these phonemes into words using a PHONEMIC DICTIONARY (a dictionary mapping words to their various possible phonemic pronunciations).

4.0.1   FEATURE EXTRACTION

Although certain aspects of speech can be identified directly from this superpositional waveform representation (e.g., energy and pitch), most of the information that distinguishes phonemes from one another is found in the relative intensities of the component waveforms at their respective oscillating frequencies [Stevens, 1998]. As the oscillating frequencies of these waveforms increase, their respective amplitudes become weaker, due to glottal pulse becoming attenuated by its interaction with the vocal tract walls. In order for the information contained at these frequencies to

---

[1]Complicating things somewhat, especially in the interests of cross-disciplinarity, is that there exist several phonemic *alphabets*. These examples come from the International Phonetic Alphabet, which is perhaps the most popular generally and certainly so among linguists, but among engineers and speech technologists, alternatives are often used for practical reasons, including one associated with the popular TIMIT database and another developed at Carnegie Mellon University. While partial mappings exist between alphabets, they are not perfect. We return to this topic in Chapter 5.

[2]Or, more specifically, TRIPHONES, which are representations of phonemes within the specific contexts of their preceeding and following phonemes.

be more accurately encoded by the acoustic model, the first stage of feature extraction is typically pre-emphasis, where the signal $x[n]$ is transformed to signal $\hat{x}[n] = x[n] - \alpha x[n-1]$ for some empirically determined parameter $0.9 \leq \alpha \leq 1.0$.

In order to extract spectral information, windows of several consecutive samples must be collectively analyzed. The width of these windows must include enough consecutive samples so that they can encapsulate two complete oscillations of the lowest-frequency waveform to be considered. This requirement of the sampling rate relative to the length of the component waveforms in time is variously referenced with regard to the Nyquist rate [Jurafsky and Martin, 2009]. Since human speech mostly occupies the frequencies between 100 Hz and 10 kHz [Stevens, 1998], the minimum window length must be $\approx 2/100\text{Hz} = 0.02\text{s}$. Rarely are analysis windows wider than this, due to the non-stationary nature of the speech signal. In order for spectral features to be extracted, however, stationarity within these windows is assumed to be inviolate.

Consecutive windows do not cover mutually exclusive segments of the audio. Indeed, in typical systems, consecutive windows overlap in time over half of their lengths (e.g., if analysis windows are 0.16 s wide, each window begins 0.08 s after its predecessor). This offset accounts for rapid changes in the speech signal. Moreover, since simple segmentation of the audio signal results in abrupt cuts to the signal, which can negatively influence proper feature extraction, each window is modified so that the signal tends to 0 at the boundaries of each window according to the popular Hamming window method [Quatieri, 2002], where

$$w[i] = \begin{cases} 0.54 - 0.46 \cos\left(\frac{2\pi i}{N}\right) & 0 \leq i \leq N-1 \\ 0 & \text{otherwise} \end{cases} \tag{4.2}$$

where N is the number of discrete samples in the analysis window. The discrete Fourier transform then converts this amplitude/time representation to its associated amplitude/frequency representation over the pre-emphasized windows obtained in this manner, $w[n]$,

$$X[k] = \sum_{n=0}^{N-1} w[n] e^{\frac{-j2\pi nk}{N}}, 0 \leq k \leq N. \tag{4.3}$$

This transform depends on Euler's formula $e^{j\theta} = \cos\theta + j\sin\theta$ for imaginary unit j. Here, the discrete Fourier transform is computed by the fast Fourier transform (FFT) algorithm using a recursive decimation-in-time[3] algorithm that forcibly assumes that N is a power of 2. The FFT computes the frequencies $X[k]$ of a signal in $O(N \lg N)$ time complexity, improving on $O(N^2)$ complexity of linear sequential computation. These frequencies are then analyzed within perceptually motivated models that imitate the behavior of the human cochlea (and the neuronal membrane therein) by means of non-linear scaling functions that warp signal frequencies f to

---

[3]Decimation in time refers to splitting into sums over even and odd time indices for the purposes of recursion.

be more amenable to feature extraction [Huang et al., 2001]. Occasionally, the Bark scale is used

$$Bark(f) = 13 \arctan(7.6 \times 10^{-4} f) + 3.5 \arctan\left(\left(\frac{f}{7.5E3}\right)^2\right), \tag{4.4}$$

but in general the spectra resultant from FFT are scaled according to the mel scale. A mel is a unit of pitch that describes the distance between sounds adjacent in their perceptual pitch [Stevens et al., 1937]. Frequencies obtained by FFT below around 1,000 Hz are mapped linearly to the mel scale, and those above 1,000 Hz are mapped logarithmically according to

$$mel(f) = 1127 \ln\left(1 + \frac{f}{700}\right). \tag{4.5}$$

This scale is analogous to the human loss of sensitivity to pitch differences at higher frequencies. The mel-scaled spectrum contains several aspects of the glottal source of speech that are not particularly useful in distinguishing between phonemes. For example, the spectrum includes fundamental frequency and energy information, which is not as important to speech recognition as details of the filter, i.e., the vocal tract. In order to deconvolve the source from the filter, the first step is to take the logarithm of the magnitude spectrum obtained in Equation 4.5. The final step visualizes this log spectrum as if it were itself a waveform and takes into consideration that the shape of this pseudo-waveform is characterized by high-frequency oscillations caused by the fundamental frequency, and otherwise by broad peaks and valleys. It is these high-frequency oscillations that correspond to the glottal source, and these broad peaks and valleys that correspond to the shape of the vocal tract (and formant frequencies, generally). Taking the spectrum of the log spectrum separates these two components and has the added benefit that the resulting coefficients are uncorrelated (unlike the spectrum), so that acoustic models used in classification do not have to encode covariances between all features, which reduces the number of parameters necessary in machine learning [Jurafsky and Martin, 2009]. This spectrum of the log spectrum, or "cepstrum", is converted from the windowed speech $w[n]$ to Mel-scaled cepstral coefficients $c[k]$ by

$$c[k] = \sum_{n=0}^{N-1} \log\left(H_k(n) \left|\sum_{l=0}^{L-1} w[l] e^{-j\frac{2\pi}{L}nl}\right|\right) e^{j\frac{2\pi}{N}kn} \tag{4.6}$$

where $H_k(n)$ is the magnitude of the mth filterbank evaluated at the nth linear frequency.

4.0.2   LINEAR PREDICTIVE CODING (LPC)

Linear Predictive Coding (LPC, also known as autoregressive modeling) estimates the main features of speech using filters $H(z)$ where

$$H(z) = \frac{X(z)}{E(z)} = \frac{1}{1 - \sum_{k=1}^{p} a_k z^{-k}} = \frac{1}{A(z)}, \tag{4.7}$$

and $A(z)$ is the inverse filter and $X(z)$ is the z-transform,

$$X(z) = \sum_{n=0}^{\infty} a^n z^{-n} = \frac{1}{1 - az^{-1}}. \tag{4.8}$$

LPC p^{th}-order analysis then predicts the current sample as a linear combination of its past p samples:

$$\tilde{x}[n] = \sum_{k=1}^{p} a_k x[n-k]. \tag{4.9}$$

Formant candidates in the spectrum can be obtained in each frame by computing the roots of the pth-order LPC polynomial $A(z)$ [Ahadi-Sarkani, 1996] using such methods as Laguerre's method. The ith root can be represented by

$$z_i = \exp\left(-\pi b_i + j 2\pi f_i\right) \tag{4.10}$$

giving the formant frequency (f_i) and bandwidth (b_i). In order to track formants over time and to decide between many formant candidates, a dynamic programming algorithm has historically been combined with an *a priori* state-based formant transition model [Huang et al., 2001], although modern alternatives exist.

4.0.3 HIDDEN MARKOV MODELS (HMMS)

A more popular alternative classification mechanism, hidden Markov models (HMMs), categorizes observable temporal data sequences according to "hidden" statistical parameterizations and an underlying connected-state structure. In speech recognition, *continuous* HMMs are defined by a multi-dimensional continuous observation space O with **o** being a sequence of length T of observation vectors $o_t \in O$ for $t = 1 \ldots T$,[4] a state space Q (where q_t is the state at time t), an initial state distribution $\pi_i = P(q_0 = i)$, a state transition matrix $A(q_i, q_j)$ describing the *a priori* probability of transitioning from state q_i to q_j, and a distribution $B_i(o)$ defining the probability of observing vector **o** in state i. Historically in ASR, the distribution $B_i(o)$ will be a mixture of Gaussians, i.e.,

$$B_i(o) = \sum_{m=1}^{M} \omega_{i,m} \frac{1}{(2\pi)^{d/2} |\Sigma_{i,m}|^{1/2}} \exp\left[-\frac{1}{2}(o - \mu_{i,m})^{\top} \Sigma_{i,m}^{-1}(o - \mu_{i,m})\right] \tag{4.11}$$

where d is the number of dimensions in each observation, $|\Sigma|$ is the determinant of Σ, and there are M component Gaussians in each state, $\omega_{i,m}$ is the weight of the mth Gaussian in state i, $\mu_{i,m}$ is its mean, and $\Sigma_{i,m}$ is its covariance.[5]

---

[4]This is analogous to the observation alphabet in discrete HMMs.

[5]"Hybrid" models combine HMMs in terms of the state structure, but replace Gaussians with neural networks, which are increasingly being used in speech technology generally.

The complete parameter set of an HMM, Φ, constitutes all parameters of $a_{ij} = A(q_i, q_j)$, $B_i(o)$, and π_i. In some cases, we are interested in computing the likelihood of a particular observation o given the parameters Φ. This is performed by the Forward algorithm [Huang et al., 2001],

$$P(\mathbf{o}; \Phi) = \sum_{\forall \mathbf{q}} P(\mathbf{q}; \Phi) P(\mathbf{o} \,|\, \mathbf{q}; \Phi), \qquad (4.12)$$

which sums over all possible sequences of hidden states $\mathbf{q}$. Here, the probability of a particular state sequence and the probability of an observation given that state sequence are

$$P(\mathbf{q}; \Phi) = P(q_1; \Phi) \prod_{t=2}^{T} P(q_t \,|\, q_{t-1}; \Phi)$$

$$P(\mathbf{o} \,|\, \mathbf{q}; \Phi) = \prod_{t=1}^{T} P(o_t \,|\, q_t; \Phi). \qquad (4.13)$$

More typically, we are interested in finding the state sequence that gives the highest probability given an observation sequence. This is referred to as "decoding" and is useful since the state sequence reveals the most likely phoneme or word sequence, assuming that individual smaller HMMs, each representing a word or phoneme, are concatenated together with appropriate bigram state transition probabilities. The Viterbi algorithm is used in decoding and determines the most likely state sequence to represent an observation, given an HMM's parameters. This algorithm can be generalized to produce a ranking of the n state sequences that give the highest probability given the observation (i.e., an "N-best list"). For the purposes of computational parsimony, this list is often approximated by generalizing the Viterbi algorithm to withhold only a limited number of state sequences up to a particular time. This is often referred to as "beam search".

In order for HMMs to model phonetic rather than lexical acoustics, three-state left-to-right structures are usually used, where the probabilities of $B(o, s)$ are modeled by multivariate GMMs.[6]

---

[6]Other probabilistic alternatives are discussed later.

CHAPTER 5

Speech Synthesis

For many, speech input to computer systems will be either impossible or at least very difficult.[1] These individuals may benefit from systems that accept text directly (Chapter 9), or optimized symbol sequences (Section 9.1), and produce synthetic speech output so that they can be heard by others. While merely useful in many public scenarios, this can be necessary for communication over the phone. Since synthesizing artificial but *intelligible* speech from text is so central to these systems, this chapter covers relevant topics in this area.

SPEECH SYNTHESIS today is normally done in software, but it was not always that way. In the 1930s, the first electronic speech synthesizer, VOCODER, was produced by Homer Dudley (and his team) at Bell Labs and consisted of a noise source and electronic oscillators that were shaped by filters controlled by foot pedals to produce vowels, consonants, and inflections that could be transmitted over a transmission wire or channel. This led to a number of algorithmic implementations of speech and signal synthesis from the same overall architecture that is still in use today.

In particular, the VOCODER system led to the first of three broad, modern categorical architectures in TEXT-TO-SPEECH (TTS) systems, delineated by their high-level methodologies in the generation of speech-like signals. Those categories are:

Formant synthesis An approach that synthesizes acoustics and formants based on rules and filters. This is a direct descendent of the work of Dudley at Bell Labs. It involves storing a small number of parameters such as formant frequencies and bandwidths for vowels, lengths of sonorants in time, and periodicity of the fundamental frequency. It can result in highly intelligible speech and is computationally inexpensive, but it can produce unnatural, "robotic"-sounding speech.

Concatenative synthesis The use of databases of stored speech to assemble new utterances. This involves selecting short sections of recorded human speech and concatenating them together in time. While the result is relatively human-like, which leads to its widespread adoption in modern commercial systems such as Apple's Siri and Microsoft's Cortana, this method usually requires a relatively large database and careful consideration when concatenating phones to avoid inappropriate blending or "glitches". Large companies can afford the resources required to build a single representative voice, which does not really scale to smaller groups building more personalized voices.

---

[1]See Chapters 7 or 8 for examples.

Articulatory synthesis The modeling of the movements of the articulators and the acoustics of the vocal tract. This often involves the uniform tube model or some other biologically inspired model of air propagation through the vocal tract. This method is computationally inexpensive, it allows us to study speech production scientifically, and it can account for particular articulatory constraints. However, the resulting speech tends to sound unnatural, and it can be difficult to adapt these systems to imitate new synthetic speakers, or even complex articulatory dynamics.

Statistical parametric synthesis Here, speech is represented using models that can also be used in speech recognition. Popular methods use hidden Markov models [Zen et al., 2009] or neural networks [Zen et al., 2013]. Methods to learn and use the former are overviewed, generally, in Section 4.0.3.

The latter two of these are discussed in more detail in Section 5.1.1.

While these three architectures differ fundamentally in their approach to speech signal generation, their use necessitates algorithmic components common to all three. Since all approaches can be used to generate speech from given electronic text that itself can be ambiguous, they must employ some degree of text analysis which includes normalization (e.g., word tokenization), homograph[2] disambiguation, grapheme-to-phoneme (i.e., letter-to-sound) rules, and the modification of prosody.

The first step in TTS tends to be simple dictionary lookup, where a provided word token is located in a PHONEMIC DICTIONARY, which maps word orthographies to sequences of phonemes used to say them. The Carnegie Mellon dictionary, for instance, has approximately 127,000 words defined as sequences of phonemes with emphasis markers (1 and 0) on the component vowels, as in:

SEWER S UW1 ER0

SEWERAGE S UW1 ER0 IH0 JH

SEWERS S UW1 ER0 Z

SEWING S OW1 IH0 NG

SEWN S OW1 N

Unfortunately, since these dictionaries are often "hand crafted", they tend to have pronunciations only in a single language or dialect (often American English). Some do not include syllable boundaries for timing or parts-of-speech, although the relatively popular UNISYN dictionary, with about 110,000 words, does include syllables, stress, and some morphology. Dictionaries alone do not typically allow for the disambiguation of homographs, which is important, since different pronunciations of word tokens massively modulate their meaning. Consider the word

---

[2]Homographs are words with the same spelling but different pronunciations, as in the emphases placed on different syllables of the word token "content" in *I am content with the content.*

token "1867", which should be pronounced *eighteen sixty seven* in reference to the year, *one eight six seven* in reference to the last four digits of a phone number, and *one thousand eight hundred and sixty seven* in reference to a quantifier, such as a dollar amount. If the TTS selects the wrong word-form expansion or prosodic emphasis, the listener will become confused or, at the least, chafed. In many cases, though not all, disambiguation can be performed simply by determining the word "class". For example, the token "use" is pronounced */y uw z/* if it is decided to be a verb, and */y uw s/* if it is a noun.

Another issue with phonemic dictionaries is that they, obviously, do not encode pronunciations for words that they do not contain. This can be a challenge—one study showed that of the 39,923 tokens in the Penn Treebank, 1775 (4.4%, 943 unique word types) were not in a typical phonemic dictionary. These tended to be names. It is therefore necessary to be able to generate phoneme sequences without using the dictionary. For this, systems use various letter-to-sound systems in which previously unseen sequences of letters (e.g., "quinquagenarianish") can be pronounced using known mappings of letters to sounds, often obtained through latent-space training such as expectation-maximization.

Finally, once a sequence of phonemes have been determined, and the appropriate locations for emphasis marked, the TTS system must decide on *how* to modulate those emphases. Prosody is the modification of speech acoustics in order to convey some extra-lexical meaning, specifically by one or more of:

Pitch Changing of the fundamental frequency or perceived frequeny over time.

Duration Modifying the lengths of sonorants (i.e., any sustained phoneme in which the glottis is vibrating).

Loudness Modifying the amount of energy produced by the lungs, or amplitude in a signal.

Clearly, placing emphasis on the right words can drastically change the meaning of a sentence. Consider the utterance *I never said she stole my money*, which carries very different meanings depending on which unique word carries the emphasis. For example, emphasizing the "I" implies that someone else claimed that she stole the money, whereas emphasizing the "money" implies that she stole something other than money. Indeed, emphases are typically placed on words in order to indicate to the listener that they are important to the meaning of the utterance, or that they encode some information thought to be new to or desired by the listener. Indeed, there are three general ways in which prosody can be used or modified in TTS systems:

Prominence Some syllables or words are more important than others, especially content words.

Structure Sentences have inherent prosodic structure. Some words group naturally together, others require a noticeable disjunction, such as at those two commas you just read.

Tune Many languages have an intonational or "sing songy" aspect to them. The perceived "musicality" of a language lends a great deal to its believability in TTS systems.

In addition to modifying a speech signal according to prominence, structure, and tune, a major component of speech technologies for people with disorders of speech production is the resulting *intelligibility* of those systems, i.e., the proportion of words that can actually be understood. Sometimes, generating speech from text is combined with methods and algorithms for *transforming* those signals themselves.

5.1   SPEECH TRANSFORMATION

We often modify an input signal $x(t)$ (or its spectral envelope $X(f)$) into an output signal $y(t)$ (or its spectral envelope $Y(f)$) by means of a transfer function $H(\cdot)$. This transfer function can operate on several domains, such as short-term frequency characteristics of a signal. In Equation 4.3 of Section 4.0.1 we introduced the Fourier transform, which determines the amplitude of an arbitrary component sinusoid with frequency f in a signal. However, those component sinusoids can also be described in terms of their *phases* which are their respective offsets in time. In this chapter, the component phases are important in the accurate definition of transformation functions $H(\cdot)$.

In order to obtain a more general parameterization of a signal to include its component phases, we define the two-dimensional complex space $s = \sigma + j2\pi f$, where f is the frequency as before, σ is the phase, and j is the imaginary unit $j = \sqrt{-1}$. The Laplace transform generalizes the Fourier transform for continuous signals as

$$X(s) = \int_{t=-\infty}^{\infty} x(t)e^{-st}\delta t. \tag{5.1}$$

Given the Laplace transform of a signal and the complex space s, a transfer function relating an input signal $X(s)$ and output $Y(s)$ is

$$H(s) = \frac{Y(s)}{X(s)} = \frac{\sum_{k=0}^{M} b_k s^k}{\sum_{k=0}^{N} a_k s^k} \tag{5.2}$$

where the roots of the numerator and denominator polynomials are the *zeros* and *poles* of the signal, respectively, and N and M are arbitrary orders of those polynomials[3] [O'Shaughnessy, 2000]. When given only a discrete sampling of the signal, Laplace is replaced by the z-transform

$$X[z] = \sum_{n=-\infty}^{\infty} x[n]z^{-n} \tag{5.3}$$

where z is a complex frequency variable analogous to s in Equation 5.1. Since Equation 5.3 only sums to a finite value on circles in the complex domain, it is normally described in polar coordinates $z = \|z\| \exp(j2\pi f/F_S)$ where F_S is the sampling frequency [O'Shaughnessy, 2000].

---

[3]An "all-pole" model defines only coefficients in the denominator.

5.1.1   CONCATENATIVE AND ARTICULATORY SYNTHESIS

In order to produce speech that is as human-like as possible, a naïve approach is to record, split, and re-assemble actual human utterances. While individual segments can be highly intelligible and natural, their concatenation often results in an unnatural-sounding discord, especially when adjacent phonemes are incompatible [Huang et al., 2001]. To avoid discontinuities of this type, vast corpora of speech segments reflecting various phonetic and emotional contexts are often stored in order to maintain some continuity. Moreover, the boundaries between sonorants are often blended using a technique called time-domain pitch-synchronous overlap-add [Moulines and Charpentier, 1990] in which signals are reconstructed by positioning adjacent segments so that they overlap according to the estimated glottal closure [Schroeter, 2008].

The vocal tract is often modeled as a concatenation of many idealized cylindrical tubes aligned at their centers where the kth tube has a cross-sectional area of A_k, as shown in Figure 5.1. Here, the volume beyond the lips is typically modeled as a tube with an infinite width. In the simplest realization of this model, the glottis produces an oscillating volume velocity, $u_G(t)$ as a function of time t, such as the spline.[4] Here, the wave produced by the glottis is often assumed to be planar and propagated along the axis of the tubes without loss due to viscosity or thermal conduction along the walls[5] [Huang et al., 2001]. If the area of a tube A is fixed, $\rho \approx 1.2$ kg/m³ is the density of air, and $c \approx 344$ m/s is the speed of sound in a human mouth, the sound waves in this model satisfy

$$-\frac{\delta p(x,t)}{\delta x} = \frac{\rho}{A}\frac{\delta u(x,t)}{\delta t}$$
$$-\frac{\delta u(x,t)}{\delta x} = \frac{A}{\rho c^2}\frac{\delta p(x,t)}{\delta t} \tag{5.4}$$

where $u(x,t)$ and $p(x,t)$ are the volume velocity (in m/s) and pressure (in kg/m³) at position x (the glottis is the origin) in the tube at time t [Quatieri, 2002]. The pressure and volume of the kth tube is then

$$u_k(x,t) = u_k^+(t - x/c) - u_k^-(t + x/c)$$
$$p_k(x,t) = \frac{\rho c}{A_k}\left[u_k^+(t - x/c) - u_k^-(t + x/c)\right] \tag{5.5}$$

where $u_k^+(\cdot)$ and $u_k^-(\cdot)$ are are the waves traveling toward the lips and glottis, respectively, and x is measured from the left-most point in the kth tube [Huang et al., 2001]. The shaping of the sound spectrum occurs because of the changes in the areas of adjacent tubes. At the junction between the kth and $k + 1^{st}$ tubes, part of the outgoing wave is reflected back into its originating tube by the reflection coefficient

$$r_k = \frac{A_{k+1} - A_k}{A_{k+1} + A_k} \tag{5.6}$$

---

[4]Generally, a spline is a piecewise function composed of polynomials.
[5]Not all models are so naïve. The Hagen-Poiseuille flow model, for instance, assumes a parabolic acoustic wave whose velocity is maximal at the axis in the direction of motion and zero at the walls [Boersma, 1998].

with larger differences in tube areas reflecting more energy [Deller et al., 2000]. The transfer function between the z-transforms of the wave velocities at the lips u_L and the glottis u_G given N concatenated idealized tubes is

$$V(z) = \frac{U_L(z)}{U_G(z)} = \frac{0.5z^{-N/2}(1+r_G)\prod_{k=1}^{N}(1+r_k)}{[1-r_G]\left(\prod_{k=1}^{N}\begin{bmatrix} 1 & -r_k \\ -r_k z^{-1} & z^{-1} \end{bmatrix}\right)\begin{bmatrix} 1 \\ 0 \end{bmatrix}} \qquad (5.7)$$

where r_G and $r_N = r_L$ are the reflection coefficients of the glottis and lips, respectively [Deller et al., 2000]. In practice, these reflection coefficients are functions of frequency so, for example, $r_L = 1$ when measuring lower frequencies of z so that all energy is transmitted, but $r_L < 1$ at higher frequencies [Huang et al., 2001]. In this model, Equation 5.7 describes the spectrum of speech at the lips given knowledge of the produced and reflected waves in the leftmost tube. To account for non-sonorant phonemes such as plosives, fricatives, and affricates, the glottal pulse train is typically replaced with a low-amplitude white-noise generator whose signal passes through $V(z)$ as before [Quatieri, 2002]. Extensions exist that allow for nasals by introducing a three-way boundary at a tube midway along the simulated vocal tract to emulate the lowering of the velum (usually with an associated closing of the rightmost tube representing the lips) [Boersma, 1998, Huang et al., 2001].

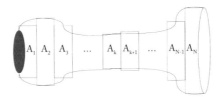

Figure 5.1: Acoustical uniform-tube model of speech production.

5.1.2   DYNAMIC MODELS OF ARTICULATION

Over the past several decades, a number of instantiations of the basic principles outlined above have arisen that implement control by an active speaker of the basic uniform-tube model through the exertion of simulated muscles. Coker [1968] proposed a model that shaped the uniform-tube model to a more realistic arrangement mimicking the midsagittal plane of a human vocal tract. Crucially, this allowed for the articulators to be explicitly identified and moved during synthesis, so that particular phonemes could be associated with desired configurations of these articulators, which were described in a code book [Coker, 1976]. The configurations of these articulators would warp the physical model of the uniform tube along horizontal, vertical, and radial dimensions, as indicated in Figure 5.2. The timing and motion between articulatory positions were in some cases explicitly defined and in others interpolated automatically through simple linear transformations.

This model is in almost all important respects reflected in the independently developed model of Mermelstein [1973], which was later extended by Rubin et al. [1981] and would come to be known as the CASY (Configurable Articulatory Synthesis) model [Iskarous et al., 2003] used in the TADA system, for instance. Other re-implementations of this same underlying model would later include low-level compensation for allophones in different articulatory contexts [Maeda, 1990], encoding of myoelastic effects between artificial muscles and their elastic connected tissues by means of spring-mass systems [Boersma, 1998], and a full complement of articulatory muscles that deform the vocal tract tube model, including the stylo-, genio-, and hyo-glossus tongue muscles and sternohyoid, for instance [Boersma, 1999].

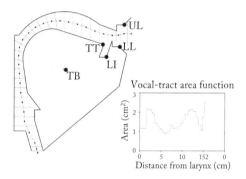

Figure 5.2: Coker model of the vocal tract for speech synthesis. Movable articulators include the tongue body (TB), tongue tip (TT), lower incisor (LI), and upper and lower lips (UL, LL), which move in tandem. Articulators move in dimensions determined by the shape of the vocal tract. Based on Boersma [1998].

5.1.3   THE KLATT SYNTHESIZER

Synthesis-by-rule is a less biologically plausible approach to speech synthesis that nevertheless focuses on the realistic acoustic properties of the generated speech. Here, a formant resonance can be generated at a specified frequency F_i and bandwidth B_i with

$$H_i(z) = \frac{1}{1 - 2e^{-\pi B_i/s_r} \cos(2\pi F_i/s_r)z^{-1} + e^{-2\pi B_i/s_r} z^{-2}} \tag{5.8}$$

where s_r is the sampling rate [Huang et al., 2001]. Klatt [1980] proposes a model which independently simulates acoustic resonances of this type given parameters determined "by hand" for various parts of speech. For vowels, a bank of six of these resonators is activated in parallel and their outputs are summed together. For nasals, similar resonances are summed together, although the zeros between resonances are also specified [McLennan, 2000]. This basic approach is one of the most popular in rule-based synthesis, and a number of derivative implementations have refined the specification of parameter values according to human data [O'Shaughnessy, 2000]. In particular, in the following sections we assume the formant parameters for frequency and bandwidth for a stereotypical male speaker as determined by Allen et al. [1987].

5.1.4 MEASURING INTELLIGIBILITY

The intelligibility of both purely synthetic and modified speech signals can be measured objectively by simply having a set of participants transcribe what they hear from a selection of word, phrase, or sentence prompts [Spiegel et al., 1990], although no single standard has emerged as pre-eminent [Schroeter, 2008]. Occasionally, ASR systems are used to approximate intelligibility. Hustad [2006] suggests that orthographic transcriptions provide a more accurate predictor of intelligibility among dysarthric speakers than the more subjective estimates used in clinical settings, e.g., Enderby [1983] and Yorkston and Beukelman [1981]. That study had 80 listeners who transcribed audio (which is atypically large for this task). It showed that intelligibility increases from 61.9% given only acoustic stimuli to 66.75% given audiovisual stimuli on the transcription task in normal speech. In the current work, we modify only the acoustics of dysarthric speech; however future work might consider how to prompt listeners in a more multimodal context.

5.1.5 ACOUSTIC TRANSFORMATION

Kain et al. [2007] propose the voice transformation system shown in Figure 5.3. This system produces output speech by concatenating together original unvoiced segments with synthesized voiced segments that consist of a summation of the original high-bandwidth signal with synthesized low-bandwidth formants. These synthesized formants are produced by modifications to input energy, F0 generation, and formant modifications. Modifications to energy and formants are performed by Gaussian mixture mapping, as described below, in which learned relationships between dysarthric and target acoustics are used to produce output closer to the target space. This process was intended to be automated, but Kain et al. [2007] performed extensive hand-tuning and manually identified formants in the input. This will obviously be impossible in a real-time system, but these processes can to some extent be automated. For example, voicing boundaries can be identified by the weighted combination of various acoustic features (e.g., energy, zero-crossing rate, first LPC coefficient) [Hess, 2008, Kida and Kawahara, 2005], and formants can be identified by the Burg algorithm [Press et al., 1992] or through simple LPC analysis (see Section 4.0.2) with continuity constraints on the identified resonances between adjacent frames [O'Shaughnessy, 2008].

Spectral modifications traditionally involve spectral filtering or amplification methods such as spectral subtraction or harmonic filtering [O'Shaughnessy, 2000], but these are not useful for dealing with more serious mispronunciations (e.g., /t/ for /n/). Hosom et al. [2003] show that Gaussian mixture mapping can be used to transform from one set of spectral acoustic features to another space. During analysis, context-independent frames of speech are analyzed for bark-scaled energy and their 24th order cepstral coefficients with

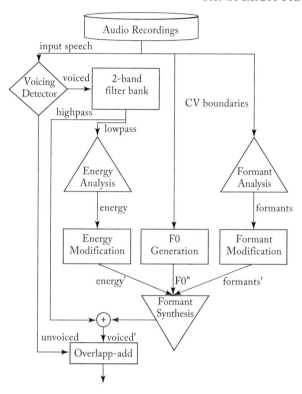

Figure 5.3: Voice transformation system proposed by Kain. Based on Kain et al. [2007].

$$X(k) = \sum_{n=0}^{N-1} x[n]e^{-j\frac{2\pi}{N}kn}$$

$$\hat{X}(k) = \log|X(Bark(k))| \qquad (5.9)$$

$$c[n] = \frac{1}{N} \sum_{k=0}^{N-1} \hat{X}(k)e^{j\frac{2\pi}{N}kn}.$$

For synthesis, a cepstral analysis approximates the original spectrum, and a high-order LPC filter is applied to each frame, and excited by impulses or white noise (for voiced and unvoiced segments). Hosom et al. show that given 99% human accuracy in recognizing normal speech data, this method of reconstruction gives 93% accuracy on the same data. They then trained a transformative model between dysarthric and regular speech using aligned, phoneme-annotated, and orthographically identical sentences spoken by dysarthric and regular speakers, and a Gaussian Mixture Model (GMM) to model the probability distribution of the dysarthric source spectral features x as the sum of D normal distributions with mean vector μ, diagonal covariance matrix Σ, and prior probability α:

$$p(x) = \sum_{d=1}^{D} \alpha_d \mathbf{N}\left(x; \mu_d, \Sigma_d\right). \tag{5.10}$$

PART II

Neurology, Anatomy, and a Few Typical Disorders

CHAPTER 6

Physical and Cognitive Foundations of Speech

When designing tools for specific speech pathologies, it can be very useful to understand some basics of how speech is produced and received both physically and in the brain. In this chapter we introduce speech production and perception generally (and very briefly), in terms of their physical biomechanics and neural aspects.

6.1 THE NEURAL ORIGINS OF SPEECH PRODUCTION

Speech and language processing recruit many parts of the brain, and a complete overview is beyond the scope of this chapter and book. However, two regions are especially relevant to certain linguistic functions and, as such, are crucial to the study of language disfunction. Indeed, observing how closed systems *fail* can be a valuable method in discovering how those systems actually *work* in the first place. The neuro-origins of speech production constitute one such closed system. Paul Broca famously made headway (so to speak) in the study of that closed system by observing a patient who could only utter a few words, which is a condition often called "expressive aphasia" (see Section 7). Broca discovered that this individual had a clear lesion in the left ventro-posterior frontal lobe, apparently caused by syphilis. This was the first direct evidence that confirmed his earlier hypothesis that language function was localized, that a particular brain structure was doing the talking. This hinted at a mechanistic view of speech production. That region, now called "Broca's area", is outlined generally in Figure 6.1, but is specifically bound to the pars opercularis and pars triangularis of the inferior frontal gyrus.

Broca's area is involved in several aspects of language function. Although historically associated with speech production, this area is also recruited in language comprehension tasks, and even in other perceptual/cognitive tasks such as the recognition of actions by others. Broca's area is connected to Wernicke's area by the arcuate fasciculus (Latin, *curved bundle*) whose importance and size appears unique to humans. Damage to this area can result in *conduction aphasia* in which patients are unable to repeat unfamiliar words, suggesting its role in short-term memory aspects of language. Wernicke's area is located in the posterior superior temporal gyrus, most commonly (as with Broca's area) in the left hemisphere. However, the location of the precise area is somewhat controversial, with some occasional overlap with the nearby auditory cortex, whose function is mostly related to acoustic processing. Given the proximiity of Wernicke's area to the auditory

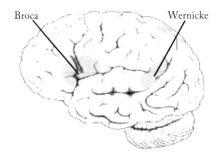

Figure 6.1: Broca's area and Wernicke's area in the left temporal lobe.

cortex, it may not be surprising that it is highly related to sensory processing, but it also appears to be involved in some resolution of word meaning, in particular for ambiguous words.

The speech musculature is controlled by the brain, where voluntary movement is initiated by the MOTOR CORTEX. Messages produced by the higher structures are transmitted through highly specialized cranial nerves (CN) that emerge through fissures in the lower brain around the cerebellum and basal ganglia. Figure 6.2 shows the cranial nerves. These nerves carry the impulses that constrict the musculature but also communicate sensory data back to the brain. All of the facial musculature is innervated by the primary facial nerve (CN VII), although submandibular and sublingual motion is also controlled by the intermediate facial nerve, and the muscles of mastication are controlled by the trigeminal nerve (CN V). Perhaps most important is the hypoglossal nerve (CN XII) which controls almost all intrinsic and extrinsic muscles of the tongue. When these cranial nerves are disrupted or rendered inoperative, partial paralysis of the respective musculature occurs as signal information is not transmitted. By contrast, if these nerves are activated involuntarily, the muscles can react in relatively unpredictable ways. Specific effects of damage to the cranial nerves are discussed in Section 8 and modeled loosely in Section 8.1.7.

6.2 THE MUSCLES OF SPEECH

Most of the organs used in speech have developed for purposes other than speaking, such as breathing or eating. Only relatively recently have these organs been adapted to speech [O'Shaughnessy, 2000]. The speech organs can be subdivided into three groups: the lungs, the larynx, and the upper vocal tract which itself consists of the jaw, lips, tongue, and nasal cavity. The lungs provide all of the airflow that is transformed by the rest of the vocal tract into the time-varying air pressure waves that constitute speech. During speech, the diaphragm muscle compresses the lungs, producing a pressure of 10–20 cm H_2O, which is about twice the amount required for normal breathing [O'Shaughnessy, 2000]. Normal breathing is generally almost inaudible since the air pressure expelled by the lungs is unobstructed by the vocal tract. Many animals, however, can create vocal noise by obstructing this air flow by means of the larynx, which is

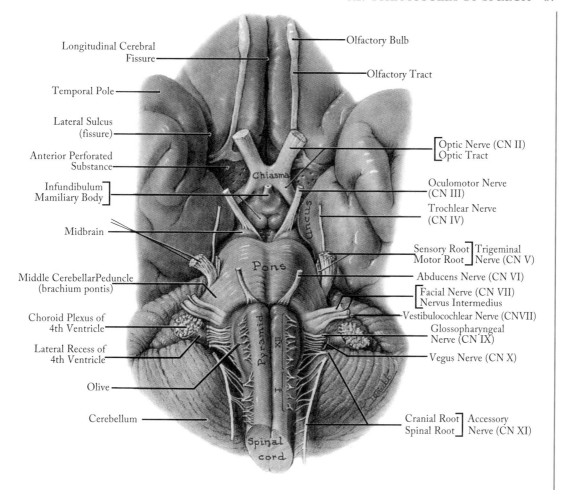

Figure 6.2: Inferior ventral view of the brain highlighting the cranial nerves, in yellow. From Moore and Dalley [2005].

typically supported by the thyroid, cricoid, arytenoid, and epiglottal cartilages [Sundberg, 1977]. These cartilages are shown in Figure 6.3, with the cricoid cartilage below the vocal fold, and the arytenoid cartilages in the posterior section which can move to abduct or adduct the vocal folds. The vocal folds do not follow muscle contractions directly, but certain muscles are involved in changing the characteristics of the quasi-periodic airflow. Changes to the fundamental frequency (F0) of speech are primarily caused by two laryngeal muscles—the vocalis muscle in the vocal folds and the cricothyroid which can increase F0 by tensing and lengthening the vocal folds by up to 4 mm [Löfqvist et al., 1984]. The fundamental frequency can also be lowered by active contractions of the THYROARYTENOID and STERNOHYOID muscles [Titze, 1994]. If these muscles cannot be controlled (i.e., contracted or relaxed), the vocal folds are tightly adducted and cannot vibrate normally, resulting in harsh, irregular F0 [Schneiderman and Potter, 2002].

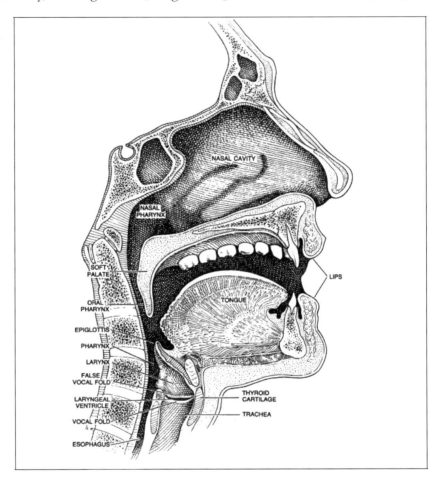

Figure 6.3: The vocal organs, as shown in the midsagittal plane. Illustration by Laszlo Kubinyi.

The muscular and bony tissue structures above the larynx contribute to speech by either warping the spectral distribution of acoustic waves or by generating certain obstruent sounds such as plosives and fricatives. The jaw is an important articulator controlled by the MASSETER and PTERYGOID muscles, although its function is largely indirect in that its placement is used to assist the positioning of the tongue and lips. The lips themselves are controlled by a number of muscles. The ORBICULARIS ORIS surrounds the mouth, protrudes the lips outward, and rounds the lips when contracted. The BUCCINATOR is a thin muscle below the cheekbones that stretches toward the mandible and controls retraction and spreading of the lips. The DEPRESSOR ANGULI ORIS and DEPRESSOR LABII INFERIORIS muscles lie below the lips and away from the midsagittal plane and pull the lip corners downward. These muscles have counterparts: LEVATOR ANGULI ORIS and LEVATOR LABII SUPERIORIS above the lips which pull the lips upward. All of these muscles are controlled by the facial nerve, described below.

The tongue is perhaps the most complex and most important articulator in speech, consisting of 12 muscle pairs and tissues [O'Shaughnessy, 2000]. The tongue provides almost all movement within the mouth, with the exception of the velum, which lowers and raises the rear of the oral cavity to allow air to pass into the nasal cavity. In normal conditions there is almost no significant lateral tongue movement, though the tongue is highly agile and can be reconfigured between relevant positions in less than 50 ms [Stevens, 1998]. The tip and dorsum of the tongue are two important areas that allow quick constrictions to occur at various positions along the vocal tract.

It is useful to conceptualize of the speech production system as a conjunction of at least two parts: a SOURCE which generates sound waves, and a FILTER which shapes those waves. In this model, the source is represented by the glottis, whose rate of vibration provides harmonics at higher multiples of that frequency. The locations of these harmonics are determined by the interaction of the sound waves with the oral cavity walls, but especially by sudden changes in the width of that cavity. These sudden changes are due predominantly to the configuration of the tongue, which is the primary causative agent of the filter. Figure 6.4 illustrates this relationship between the physical contour of the tongue and the resulting effect on the distribution of the formants of vowels.

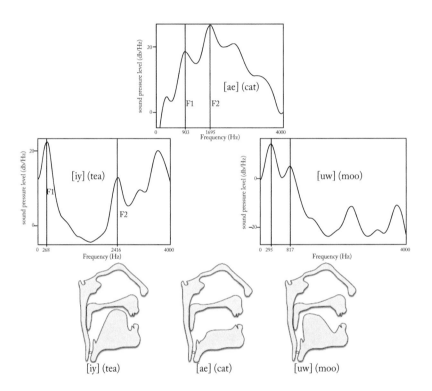

Figure 6.4: Exemplar configurations of the tongue for three English vowels and the resulting spectral envelopes. These examples demonstrate the effect that the dorsoventral (front-back) position of the tongue has on the distribution of F2 and that the superior-inferior (top-down) position has on F1. Based on Jurafsky and Martin [2009].

CHAPTER 7

Dementia and Aphasia

So far, the focus of much of this book has been on speech technologies that aid in communication for those whose acoustics of speech are unintelligible. However, individuals with cognitive differences may also benefit from technologies that either help to assemble meaning out of received syntax and semantics, or help to produce meaningful utterances of their own.

DEMENTIA, as associated with Alzheimer's disease (AD), is a progressive neurodegenerative disorder with memory impairment as the primary deficit, followed by declines in language, ability to carry out motor tasks, object recognition, and executive functioning [American Psychiatric Association, 2000]. While AD remains the most prevalent, there are several other kinds of dementia with similar though distinct symptoms:

Creutzfeldt-Jakob disease is caused by misfolded prion proteins that have a cascade effect through the brain, resulting in relatively rapid rates of fatality. It is the most common form in humans of a broader class of similar disorders across other mammals (e.g., "mad cow disease" in cattle). It is commonly associated with depression, agitation, mood swings, muscle stiffness, and spasticity.

Frontotemporal dementia encapsulates a few different types of dementia including primary progressive aphasia (PPA), Pick's disease, progressive supranuclear palsy, and others. As the name suggests, frontal and temporal regions of the brain are primarily affected, which therefore affects language. In particular, PPA affects language before other kinds of behavior and has two main variants (among others): *semantic dementia* is manifested by typical and easy fluency but with less meaningful words and phrases; *progressive nonfluent aphasia* typically makes speech more halting, tentative, and agrammatical.

Huntington's disease is caused by a single defective gene (on chromosome 4) which alters typical proteins in the brain. Unlike other forms of dementia, it typically develops at a younger age, usually between 30 and 50, although it is possible to occur at any age. Huntington's disease causes uncontrolled movement, reduced reasoning skills, and difficulty in planning, concentration, and memory. Depression and anxiety are also common, as is obsessive-compulsive disorder.

Lewy body dementia is caused by atypical aggregatins ("clumps") of the protein α-synuclein which limit pathways in the cortex generally. Symptoms of this type of dementia are similar to those in AD (i.e., memory loss and thinking problems), but are more commonly associated with sleep disturbances, vivid visual hallucinations, and physical rigidity as in Parkinson's disease.

Parkinson's disease (PD) is also very common, affecting nearly 2% of people over the age of 65. While the symptoms of PD are most typically related to motor movement, rigidity, and spasticity initially, 50 to 80% of individuals with this disorder develop dementia particular to PD over time. PD is also associated with clumps of α-synuclein, but these are typically located deeper in the brain, namely the substantia nigra, and are thought to degenerate the nerve cells that produce dopamine.

Vascular dementia accounts for approximately 10% of dementia cases and is caused by inadequate blood flow to the brain, subsequent to risks such as smoking, sedentary lifestyles, advanced age, or inappropriate levels of cholesterol and blood-sugar as the result of one's diet. Symptoms of vascular dementia may be most obvious very soon after a traumatic event such as a major stroke—these include confusion, disorientation, impaired judgement, vision loss, and, crucially, changes to language. The locations and extent of brain injury (incl. microscopic bleeding and blood-vessel blockage) naturally determine the distribution of symptoms. Typically, neuro-imaging is used in the diagnosis of vascular dementia.

Wernicke-Korsakoff syndrome is chiefly a memory disorder caused by an acute deficiency in thiamine (vitamin B1), most commonly associated with alcohol abuse. Memory loss is the most common symptom.

Dementia impacts many aspects of an individual's life, including declines in activities of daily living (ADL) such as shopping, finances, housework (called *instrumental* ADLs) and self-care tasks (called *basic* ADLs, such as bathing). Dementia is often associated with broad changes to the brain, such as the progression of AD from the memory centers, through the language centers, and onward through the cortex generally. Language changes in older adults with dementia include increasing word-finding difficulties, loss of ability to verbally express information in detail, increasing use of generic references (e.g., "it" instead of "the petunia"), and progressing difficulties understanding information presented verbally [American Psychiatric Association, 2000]. By contrast, the deterioration in language comprehension and/or production resulting from *specific* brain damage is called APHASIA, of which there are also several types:

Broca's aphasia Here, the lesion or degeneration is located around and in Broca's area, often extending backward along the Sylvian fissure. Because of its symptoms, it has sometimes been called *expressive* or *motor* aphasia. Often these symptoms are manifested by slow, halting, or hesitant speech that can appear to be the result of difficulties in word finding. Long-term suprasegmental information, such as rhythm and prosody, are also diminished. Syntax becomes simpler, as in sentences being replaced by words or short phrases, and even morphological aspects of words are lost, effectively lemmatizing many words. Comprehension, by contrast, is relatively unaffected.

Wernicke's aphasia As one might expect, Wernicke's aphasia involves damage to Wernicke's area. By contrast to Broca's aphasia, speech in Wernicke's aphasia can appear superficially to be highly fluent, with natural phrase boundaries and long-term suprasegmental prosody resembling completely natural speech. However, the difficulties tend to occur in comprehension and circumlocutions, and completely unintelligible sequences ("gobbledegook") are common. Often, the speaker can appear to be unaware that their speech is essentially meaningless, or at least wholly unintelligible to the listener. As such, it is sometimes called *receptive* or *sensory* aphasia.

Global aphasia This is a combination of Broca's and Wernicke's aphasia. Here, there can be a total reduction in all aspects of spoken and written language.

Sometimes (not always), the symptoms of aphasia (especially Broca's variant) can be indistinguishable from more global dementia—e.g., difficulties in recalling words during speech can be both caused by specific damage to Broca's area and to memory centers more broadly. It is important, clinically, to differentiate between these forms of aphasia and other linguistic disorders, including AGNOSIA, which specifically involves a difficulty in identifying sensory stimuli (including the sounds of speech in its *auditory* variant and symbols of written language in its *visual* variant) and APRAXIA, which is a difficulty in organizing speech movements in the part of the nervous system between the high-level cortices and the motor neurons that can result in severe unintelligibility, and is distinct from dysarthria and aphasia.

7.0.1 LANGUAGE USE IN DEMENTIA AND ALZHEIMER'S DISEASE

In order to design a speech interface for individuals with dementia, and AD in particular, it is important to understand how their speech differs from that of the general population. Guinn and Habash [Guinn and Habash, 2012] showed, through an analysis of conversational dialogs, that repetition, incomplete words, and paraphrasing were significant indicators of Alzheimer's disease, but several expected measures such as filler phrases, syllables per minute, and pronoun rate were not. Indeed, pauses, fillers, formulaic speech, restarts, and speech disfluencies are all hallmarks of speech in individuals with Alzheimer's [Davis and Maclagan, 2009, Snover et al., 2004]. Effects of Alzheimer's on syntax remain controversial, with some evidence that deficits in syntax or of agrammatism could be due to memory deficits in the disease [Reilly et al., 2011].

Other studies have applied similar analyses to related clinical groups. Pakhomov et al. [2010] identified several different features from the audio and corresponding transcripts of 38 patients with frontotemporal lobar degeneration (FTLD). They found that pause-to-word ratio and pronoun-to-noun ratios were especially discriminative of FTLD variants and that length, hesitancy, and agrammatism correspond to the phenomenology of FTLD. Roark et al. [2011] tested the ability of an automated classifier to distinguish patients with mild cognitive impairment from healthy controls that include acoustic features such as pause frequency and duration.

Difficulty with naming tasks is one of the first hallmarks of the disease and can occur early, as the disease progresses [Kirshner, 2012]. There is some evidence that patients with AD may have more difficulty naming verbs than nouns [Robinson et al., 1996], although the effect of AD on syntax is controversial. Some researchers have reported syntactic impairments in AD, while others claim that any apparent deficits are in fact due to difficulties with memory and semantics. Given word-finding difficulties and a reduced vocabulary, the language of AD patients can seem "empty" and "verbose and circuitous, running on with a semblance of fluency, yet incomplete and lacking coherence" [Appell et al., 1982]. Macro-linguistic language functions, such as understanding metaphor and sarcasm, also tend to deteriorate in AD [Rapp and Wild, 2011].

7.0.2   COMMUNICATION DIFFICULTIES

The trouble source-repair (TSR) model describes difficulties in speaking, hearing, or understanding, including how communication repairs are initiated and carried out [Schegloff et al., 1977]. Difficulties can be phonological (e.g., mispronunciation), morphological/syntactic (e.g., incorrect agreement among constituents), semantic (e.g., disturbances related to lexical access, word retrieval, or word use), and discourse (i.e., misunderstanding of topic, shared knowledge, or cohesion) [Orange et al., 1996]. The majority of TSR sequences involve self-correction of a speaker's own error, e.g., by repetition, elaboration, or reduction of a troublesome utterance [Schegloff et al., 1977].

Orange et al. [1996] showed that while 18% of non-AD dyad utterances involve TSR, 23.6% of early-stage AD dyads and 33% of middle-stage AD dyads involved TSR. Of these, individuals with middle-stage AD exhibited more discourse-related difficulties including inattention, failure to track propositions and thematic information, and deficits in working memory. The most common repair initiators and repairs given communication breakdown involved frequent *wh*-questions and hypotheses (e.g., "*Do you mean…?*"). Conversational partners of individuals with middle-stage AD initiated repair less frequently than conversational partners of control subjects, possibly aware of their deteriorating ability, or to avoid possible further confusion.

An alternative, though closely related, paradigm for measuring communication breakdown is trouble indicating behavior (TIB) in which the confused participant implicitly or explicitly requests aid. In a study of 7 seniors with moderate/severe dementia and 3 with mild/moderate dementia, Watson [1999] showed that there was a significant difference in TIB use ($p < 0.005$) between individuals with AD and the general population. Individuals with AD are most likely to exhibit dysfluency, lack of uptake in the dialog, metalinguistic comments (e.g., "*I can't think of the word*"), neutral requests for repetition, whereas the general population are most likely to exhibit hypothesis formation to resolve ambiguity (e.g., "*Oh, so you mean that you had a good time?*") or requests for more information. This model incorporates 12 distinct types of TIB.

1. **Neutral or non-specific requests for repetition (local)**. Minimal queries indicating non-understanding, which did not identify the problem specifically. *E.g., What? Huh?*

2. **Request for confirmation—repetition with reduction**. Partial repair of a trouble source, often in the form of a question. *E.g., Speaker 1: I went to a party last night. Speaker 2: Last night?*

3. **Request for confirmation—complete repetition**. Recapitulatory "echo" questions, often with pronoun alternation. These follow a similar pattern and therefore must be distinguished from expressions of incredulity or disapproval. *E.g., Speaker 1: I went to a party last night. Speaker 2: You went to a party last night?*

4. **Request for confirmation—repetition with elaboration**. Same as TIB 3, but with the inclusion of additional semantic content. *E.g., Speaker 1: I went to a party last night. Speaker 2: You went to a party last night …at the zoo?*

5. **Request for specific information**. Contains a specific semantic concept, content word, or referent to the previous or recent turn. *E.g., Speaker 1: I played a bit of golf. Speaker 2: What did you play a bit of?*

6. **Request for more information**. A non-specific request (i.e., without direct mention of semantic concepts in a recent utterance). The speaker did not provide the amount or type of information desired by the listener. *E.g., I don't understand. Tell me more. What do you mean?*

7. **Corrections**. Are the result of a violation in the quality of message or message inaccuracies. Here, semantic confusion often originates from the individual not indicating the TIB. *E.g., Speaker 1: It's Monday. Speaker 2: No, it's Friday.*

8. **Lack of uptake / lack of continuation**. Verbal behaviors including (i) minimal feedback where back channel responses indicate non-understanding or lack of contribution to or elaboration on topic extension; (ii) overriding where a participant does not allow the floor; and (iii) topic switch where one participant abruptly changes topic. *E.g., Speaker 1: Do you know what indexed means? Speaker 2: Yes. Speaker 1: What? Speaker 2: Oh, it's a bit too hard, bit late too late to…*

9. **Hypothesis formation**. Guessing behaviors involving supplying words or speaking for or on behalf of the other participant. This does not include hypotheses in the form of rhetorical questions (which are instead categorized as TIB 5). *E.g., Speaker 1: We went to the farm. Speaker 2: You went to Riverdale Farm.*

10. **Metalinguistic comment**. This includes "talk about talk" which explicitly refers to non-understanding of message content, the interpersonal manner in which the message was conveyed, or the production of the message. *E.g., I can't remember. I don't understand.*

11. **Reprise / minimal dysfluency**. Reprises in which partial or whole repetition or revision of the message occurs. Minimal dysfluencies indicate difficulties producing a message that involve sound, syllable and word repetition, pauses, and fillers. These are deemed more excessive than the typical dysfluencies that occur in typical speech. *E.g., Eerrr, I want to—we went to the river.*

12. **Request for repetition—global**. Minimal queries indicate a non-understanding of the previous section of talk. *E.g., Wait—go back to the part about… You just lost me.*

Automated communicative systems that are more sensitive to the emotive and mental states of their users are often more successful than more neutral conversational agents [Saini et al., 2005]. In order to be useful in practice, these systems should therefore mimic some of the techniques employed by caregivers of individuals with AD. Often, these caregivers are employed by local clinics or medical institutions and are trained by those institutions in ideal verbal *communication strategies* for use with those having dementia [Goldfarb and Pietro, 2004, Hopper, 2001]. These include [Small et al., 2003] but are not limited to:

1. Relatively slow rate of speech.
2. Verbatim repetition of misunderstood prompts.
3. Closed-ended questions (i.e., that elicit yes/no responses).
4. Simple sentences with reduced syntactic complexity.
5. Giving one question or one direction at a time.
6. Minimal use of pronouns.

Automated systems that are designed to communicate with older adults with dementia should incorporate some of these strategies in their output.

CHAPTER 8

Dysarthria

The canonical use case for much speech technology is for individuals who are cognitively typical and therefore capable of abstract *language* but whose speech *acoustics* are acutely atypical.

Dysarthria is an "umbrella" term that refers to a set of both congenital and traumatic neuromotor disorders that impair the physical production of speech. These impairments reduce or remove normal control of the primary vocal articulators but do not affect the regular comprehension or production of meaningful, syntactically correct language. Congenital causes of dysarthric speech are often manifested by some sort of asphyxiation of the brain, inhibiting normal development in the speech-motor areas. Of these causes, cerebral palsy is among the most common,[1] affecting approximately 0.5% of children in North America, 88% of whom are dysarthric throughout adulthood [Augmentative Communication Incorporated (ACI), 2007]. Later-onset causes are more typically traumatic, including cerebro-vascular stroke affecting approximately 1% of the population aged 45 to 64, and 5% of those aged 65+, with the severity of impairment varying with the amount of cerebral damage [Augmentative Communication Incorporated (ACI), 2007]. Other sources of dysarthria include multiple sclerosis, Parkinson's disease, myasthenia gravis (i.e., blocked acetylcholine receptors), and amyotrophic lateral sclerosis [Kent and Rosen, 2004].

Neurological causes of dysarthria involve damage to the cranial nerves that control the articulatory musculature of speech. For example, damage to the recurrent laryngeal nerve typically reduces control over vocal fold vibration (i.e., phonation), resulting in either guttural or grating raspiness. Inadequate control of soft palate movement caused by disruption of the vagus cranial nerve may lead to a disproportionate amount of air being released through the nose during speech (i.e., hypernasality). More commonly, a lack of tongue and lip dexterity often produces heavily slurred speech and a more diffuse and less differentiable vowel target space [Kent and Rosen, 2004]. The lack of articulatory control often leads to various involuntary sounds caused by velopharyngeal or glottal noise, or noisy swallowing problems [Rosen and Yampolsky, 2000]. Dysarthria is differentiated from *apraxia*, in which damage to Broca's area in the left frontal lobe reduces the ability to plan rather than to execute speech articulation.

Some of the more clearly defined subgroups of dysarthria include the following:

---

[1]The earliest record of a scientific understanding of cerebral palsy dates from 1861 when Dr. William John Little described a systematic condition in children characterized by "spastic rigidity of the limbs of new-born children, [and] spastic rigidity from asphyxia neo-natorum" [Little, 1861]. This condition gradually came to be known as Little's disease, later generalized to incorporate speech and swallowing difficulties [Posey, 1923], and later still redefined as spastic displegia—a type of cerebral palsy.

Spastic Due to upper motor neuron lesions, pyramidal tract damage, and especially lesions to the facial and hypoglossal cranial nerves for jaw and tongue movement, respectively. Phonation is harsh, and strained with low sustained pitch [Duffy, 2005]. Hypernasality often accompanies phonemes $/p/$, $/b/$, $/s/$, and $/k/$. Bursts of loudness, slow rate of speech, and reduced onset time distinction between voiced and unvoiced stops are also associated with spastic dysarthria [Hasegawa-Johnson et al., 2006b].

Hyperkinetic Due to lesions in basal ganglia, and often accompanies other involuntary movement. Harsh phonation is comparable to spastic dysarthria, although hypernasality is more common and involuntary movements tend to superimpose on voluntary articulations. Slowness is also common.

Hypokinetic Associated with Parkinsonism, and due to lesions in the basal ganglia, or to either anti-psychotic medication or blows to the head. Hypokinetic dysarthria results in mono-pitch hoarse phonation with very low monotonous volume. Compulsive syllabic repetition (pallilalia) can also occur. It can result in difficulty initiating voluntary speech, or sudden interruption of movement during speech [Duffy, 2005].

Ataxic Caused by damage to cerebellar control of respiration, phonation, and articulation, but is chiefly characterized by pronounced bursts of loudness. Equal and excessive stress on each spoken syllable is also common. Discoordination results in slurred and slow speech, where patients sound as if inebriated [Duffy, 2005].

Flaccid Caused by damage to the lower motor neurons. May result in complete paralysis of one or more vocal folds, causing breathiness, low volume, increased nasality, and monotonous pitch. In unilateral paralysis the jaw may deviate to the weakened side while the tongue moves toward the stronger side, sometimes resulting in drooling [Duffy, 2005].

Despite many overlapping behaviors between these categories, there may also exist some clear delineations. For example, Ozawa et al. [2001] have shown that slow speech in spastic dysarthria is more often caused by lengthened syllables, relative to ataxic dysarthria, which is categorized by longer pauses. Nishio and Niimi [2001] reach a similar conclusion, although they focus on flaccid and hypokinetic dysarthria as predictors for longer pauses. Even if invariant distinctions exist between types of dysarthria for certain features, it is not clear that prior knowledge of differing neuromotor deficiencies can be exploited in ASR.

The following subsections describe common phenomena in dysarthric speech, including abnormal speaking rates, fatigue, disfluency, and reduced control of volume, articulation, and pitch.

8.1   PRESENTATION AND ASSESSMENT

There are several ways in which dysarthria presents itself, outlined below.

8.1.1   ATYPICAL SPEAKING RATES

Dysarthric speech is often between 10 and 17 times slower than regular speech, at about 15 words per minute in the most severe cases [Patel, 1998]. Apart from being more laborious for the speaker and listener, slow speech has several acoustic consequences. For example, monosyllabic words that are prolonged by lengthened voiced phonemes (e.g., vowels) are frequently misinterpreted as multisyllabic by human listeners [Kent and Rosen, 2004]. Also, if lengthy occlusions precede voiceless plosives such as $/k/$, $/p/$, or $/t/$, listeners often mispartition a single word into two [Raghavendra et al., 2001]. Despite a great amount of inter-speaker variability, dysarthric individuals who can maintain a regular speaking rate are able to repeat individual speech units with fairly normal consistency [Kent and Rosen, 2004].

Abnormally slow speaking rates have been shown to expand the acoustic vowel space, leading to increased intraspeaker variability for those speakers, and more difficult differentiation between phonemes [Kent and Rosen, 2004]. Tsao et al. contest the significance of this vowel space expansion in general, but agree that the acoustics of speech is far more variable among slow speakers, including higher interspeaker variability within that group [Tsao et al., 2006]. Simple alterations of speaking rate alone, however, do not account for all unintelligibility of dysarthric speech [Hammen et al., 1994].

8.1.2   MUSCLE FATIGUE AND WEAKNESS

Low endurance of the facial muscles is often associated with dysarthria, and may be caused by deficiencies at different points of the neuromotor process. Reduced lip and tongue strength and tongue endurance have been associated with Parkinsonism [Solomon et al., 2000], stroke [Thompson et al., 1995], myasthenia gravis [Weijnen et al., 2000], and traumatic brain injury [Goozee et al., 2001]. Muscle weakness may also limit the amount of air these speakers can release, therefore reducing acoustic energy. Despite a clear correlation between dysarthria and facial muscle weakness [Umapathi et al., 2000], the acoustic consequences of that weakness may be less important than other features of disordered speech [Goozee et al., 2001]. McHenry and Liss [2006] suggest that temporal and spatial inconsistencies of hypokinetic and ataxic dysarthria influence acoustic perception more than increased hypernasality caused by velopharyngeal weakness.

8.1.3   INTENSE ACOUSTIC DISFLUENCY

The lack of articulatory control in dysarthria often leads to various involuntary sounds caused by velopharyngeal or glottal noise, or noisy swallowing problems [Rosen and Yampolsky, 2000]. Figure 8.1 shows examples of both involuntary noise and involuntary pausing in dysarthric vs. normal pronunciations of the word *five*. An ASR system that hasn't been optimized for this kind of speech may misinterpret the noise as phonemes, and pausing as a word boundary. This is compounded by a mispronunciation of the labiodental $/v/$ as the labial $/m/$.

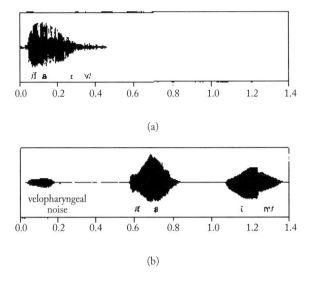

Figure 8.1: Example waveforms of typical (a) and cerebral palsied (b) pronunciation of the word *five*. Based on Chen and Kostov [1997].

Other types of disfluency commonly associated with dysarthria include hesitation (e.g., false-starts), stuttering, and other involuntary repetition, although these may sometimes result from higher-level linguistic causes [Kent, 2000]. These sorts of disfluencies produce severely atypical phrasing which is difficult to understand at the utterance level, without familiarity with the speaker.

8.1.4   REDUCED CONTROL OF ARTICULATION AND PITCH

The most common dysarthric mispronunciations tend to occur with more complex requirements on articulatory movement, namely consonants or consonant clusters. Thubthong et al. [2005] report that among 18 children with CP, word-initial consonants were the most difficult to pronounce, with only a 62.2% rate of accuracy. Of these, alveolar consonants were the most troublesome, with /r/ and /t/ being correctly articulated 0% and 27.8% of the time, respectively. Vowels and word-final consonants were the most accurately articulated phoneme classes, at 93.7% and 77.1% accuracy respectively. Groups of clustered consonants such as /tr/ or /kw/ were produced correctly only 11.1% of the time.

Figure 8.2 shows pronunciation of the word *yes* by both a control and a cerebral palsied individual. The reduced precision of the fricative /s/ is likely caused by insufficient jaw movement, and the prolonged duration is almost exclusively due to an extended vowel. Interestingly, the distribution of formants in Figure 8.2(b) suggests a pronunciation closer to /i/ [O'Shaughnessy, 2000] than to ε as in Figure 8.2(a).

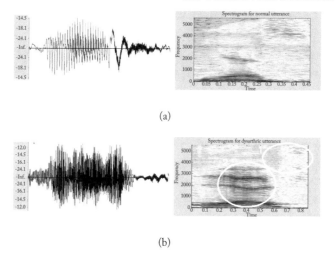

Figure 8.2: Example waveforms (dB vs. time) and spectrograms (frequency vs. time) of typical (a) and atypical (CP) (b) pronunciation of the word *yes*. Note in the latter the suppression of the /*s*/ fricative in region 1, and the misplaced formants in region 2. From Polur and Miller [2006].

Kim et al. [2010b] found that speakers with spastic cerebral palsy have drastically reduced displacement of the tongue tip, in position and in velocity. They also found that directed motion of the tongue occurred later than voicing onset relative to the general population, which loosely supports other work that suggests a general difficulty in co-ordinating glottal and supraglottal systems in dysarthria [Chen and Stevens, 2001].

8.1.5 PITCH PROSODY

Prosody includes changes in fundamental frequency F_0 caused by voluntarily tensing the vocal folds to stress or decline certain syllables for syntactic effect. Proper control of the pitch aspect of prosody has several positive effects on intelligibility, and is also an important conveyor of semantic and emotional content [O'Shaughnessy, 2000].

Dysarthria often reduces voluntary control of the larynx, reducing or diminishing prosody and resulting in machine-like speech [Mori et al., 2005]. Despite this reduction, however, dysarthric speakers retain at least the ability to reliably form differentiable questions and statements using binary vocal pitch contours [Patel, 2002a,b]. Kim et al. [2010a] suggest that dysarthric speakers use pitch and intensity cues of lexical stress to a greater degree than non-dysarthric speakers, especially in words that emphasize the second syllable. Inappropriate pitch prosody may affect up to 50% of suspected dysarthric children [Ziegler and Maassen, 2004], but causes can either be physical or learned (e.g., through compensatory behavior).

8.1.6   EVALUATING AND TREATING DYSARTHRIA

Intelligibility quantifies the degree to which an individual's speech is discernible to human listeners, typically by measuring the average accuracy of word-level transcriptions of utterances across groups of naïve listeners [Kent et al., 1989] [Hasegawa-Johnson et al., 2006a, Menendez-Pidal et al., 1996]. If speech samples are phonetically balanced, one can automatically classify the most prevalent errors according to discrete phonetic features of how the vocal tract restricts airflow (manner), where along this tract the narrowest constriction occurs (place), and whether the vocal folds vibrate during production (voicing).[2] Other procedures that measure intelligibility include the Children's Speech Intelligibility Measure which includes developmental statistics, and the Yorkston-Beukelman-Traynor assessment [Hammen et al., 1994] which has been computerized and includes factors such as speaking rate and rate of intelligibility. Intelligibility scores are also sometimes accompanied by results of the Frenchay Dysarthria Assessment [Enderby, 1983] that individually scales the strength of the various articulators, respiration, reflex, and rate [Menendez-Pidal et al., 1996].

Since dysarthrias cannot yet be cured with surgery or medication, behavioral interventions are often used to strengthen the articulatory muscles or develop alternate pronunciation strategies to improve intelligibility [Kent, 2000]. This behavioral intervention often involves computer-based treatment that can improve intelligibility by exercises and feedback automatically generated using speech recognition [Thomas-Stonell et al., 1998] that is just as effective as traditional treatment [Palmer et al., 2007].

The precise relationship between speech repeatability and neurological damage is still an open question. For instance, although dysarthric utterances are highly variable, those dysarthric speakers who have maintained a regular speaking rate appear to be able to repeat individual speech units in isolation with fairly normal reproducibility [Chen and Kostov, 1997, Kent and Rosen, 2004]. Furthermore, although intelligibility strongly correlates with recognition accuracy in ASR [Mengistu and Rudzicz, 2011], consistency does not [Thomas-Stonell et al., 1998].

8.1.7   A NOISY-CHANNEL MODEL OF DYSARTHRIA

Dysarthria is sometimes characterized as a distortion of parallel biological pathways that corrupt motor signals before execution [Freund et al., 2005, Kent and Rosen, 2004]. It may be possible to cast the interface between the origins of speech production in the cerebral cortex with the innervations of the muscles involved in speech within the framework of the noisy-channel model. If so, this would allow for motor signal distortions in dysarthria to be learned given appropriate measurements of the vocal tract.

At its simplest, the noisy channel can be described by two variables, the source signal X (what is said into a telephone at one end of a line) and the received signal Y (at the other end), cast into two probabilities, the probability of the source signal being produced in the first place, $P(X)$, and the probability that a medium corrupts X to "sound like" Y in specific ways, $P(Y \mid X)$. With

---

[2]These measures may be overly simplistic, but are useful in classification [O'Shaughnessy, 2000].

this, decoding the received signal Y (i.e., selecting your best guess, X^*, of the original source signal X) becomes

$$X^* = \arg\max_X \ P(Y \mid X) \cdot P(X).$$

This scenario is exemplified in Figure 8.3.

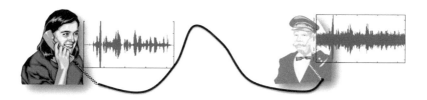

Figure 8.3: An abstract representation of the noisy channel. The producer of a message X, on the left, passes that message through a medium (in this case a telephone wire) which corrupts the signal in statistically understandable ways, producing the message Y at the receiver's end, on the right.

Although not quantified here, we detect that a lack of articulatory control can often lead to observable acoustic consequences. For example, our dysarthric data contain considerable involuntary types of velopharyngeal or glottal noise (often associated with respiration), audible swallowing, and involuntary repetition.

We have considered the amount of statistical disorder (i.e., entropy) in both acoustic and articulatory data in dysarthric and non-dysarthric speakers. The use of articulatory knowledge reduces the degree of this disorder significantly for dysarthric speakers (18.3%, relatively), though far less than for non-dysarthric speakers (86.2%, relatively). In real-world applications we are not likely to have access to measurements of the vocal tract; however, many approaches exist that estimate the configuration of the vocal tract given only acoustic data [Richmond et al., 2003, Toda et al., 2008], often to an average error of less than 1 mm. The generalizability of such work to new speakers (particularly those with dysarthria) without training is an open research question.

We have argued for noisy channel models of the neuro-motor interface assuming that the pathway of motor command to motor activity is a linear sequence of dynamics. The biological reality is much more complicated. In particular, the pathway of verbal motor commands includes several sources of sensory feedback [Seikel et al., 2005] that modulate control parameters during speech [Gracco, 1995]. These senses include exteroceptive stimuli (auditory and tactile), and interoceptive stimuli (particularly proprioception and its kinesthetic sense) [Seikel et al., 2005], the disruption of which can lead to a number of production changes. For instance, Abbs et al. [1976] showed that when conduction in the mandibular branches of the trigeminal nerve is blocked, the resulting speech has considerably more pronunciation errors, although is generally intelligible. Barlow [1989] argues that the redundancy of sensory messages provides the necessary input to the motor *planning* stage, which relates abstract goals to motor activity in the cerebellum.

As speech recognition systems are continually improved for atypical speech, including those designed for speakers with dysarthria, one potential avenue for future research involves the incorporation of feedback from the current state of the vocal tract to the motor planning phase. This would be similar, in premise, to the DIVA model [Guenther and Perkell, 2004]. There is also ample evidence for a "phonological store" or "phonological loop" in the cerebellum in which articulatory rehearsals and their expected acoustic consequences are stored for between 1.5 and 2.0 seconds during speech comprehension and production [Baddeley et al., 1998, Beaman, 2007]. Under this model, speakers with dysarthria show a normal capacity for articulatory rehearsal, which suggests that distortions occur after the planning stage but before motor execution [Baddeley and Wilson, 1985].

PART III

Technologies that Enable Expression

CHAPTER 9

Augmentative and Alternative Communication

AUGMENTATIVE AND ALTERNATIVE COMMUNICATION (AAC) encapsulates a broad collection of technologies that enable both expression and reception of language. These technologies include high-tech text-entry systems, such as specialized keyboards, that compose messages in abstracted alphabets and produce speech or text as output. These technologies are used across the spectrum of ages and speech-language impairments, from congenital issues such as cerebral palsy and autism to acquired conditions including Parkinson's disease and Alzheimer's. Additionally, there are approximately 270,000 people in North America with spinal cord injuries, approximately 47% of whom develop quadriplegia (i.e., tetraplegia), in which partial or total paralysis of the limbs and face [Spears and Holtz, 2010, Walls et al., 2009] can necessitate the use of AAC devices.

AAC devices are often prescribed by registered SPEECH-LANGUAGE PATHOLOGISTS (SLPs), who are generally represented by hospitals and clinics and are members of professional organizations (e.g., CASLPO[1]). According to the U.S. Bureau of Labor, there were 120,000 SLPs in the United States in 2008.

Currently, SLPs use very conventional methods in their interaction with clients and patients during therapy—typically no more than pen and paper. The common practice is for clients to come on site to have one-on-one sessions with their therapists and to do exercises at home. Typically, this will involve a lot of travel and the use of paper-based exercise books. This approach means that SLPs cannot guarantee that clients are performing exercises correctly at home nor can they maintain detailed charts with specific voice information.

Several companies produce portable aids for people with communication differences. These companies often share a considerable amount of underlying technology. For one, AAC devices must generally involve text input, which necessitates a simple and efficient approach to their creation. This is a topic covered next in Section 9.1 and to a large extent builds on N-gram models of word prediction covered previously in Section 3.1.

Words+ Incorporated, for example, makes hand-held products costing around $3,500 that can either replay previously recorded voice messages or synthesize artificial speech from sentences pasted together from a series of key presses. Other companies such as Dynavox[2] and Prentke

---

[1] http://www.caslpo.com/

[2] Note that Dynavox and Tobii are often discussed separately for historical reasons, although they merged in 2014 to form a single company, TobiiDynavox (http://www.tobiidynavox.com/).

Romich produce similar systems costing between $7,300 and $12,000 that consist either of alternative keyboards or "scanning" interfaces of the type used by Stephen Hawking where a button is pressed to iteratively cycle through lists of commands, words, or phrases. While popular and often advantageous, there are also a number of drawbacks to this approach. First, the vocabularies supported by these systems are limited to about 6,500 words which do not include many words specific to the user (e.g., family or place names); new words can only be added through laborious processes. Systems that do not require a pre-programmed vocabulary may be beneficial, but are rare unless modifications can be made at the acoustic level, as in recent work with the TORGO database [Rudzicz, 2013]. Second, the neurological damage that causes speech disorders usually affects other physical activity, which can drastically limit mobility. For instance, speakers with severe dysarthria can be up to 30 times slower at keyboard interaction than spoken interaction, and the former tends to be far more fatiguing [Hosom et al., 2003]. The relatively small keys used in the Word+ and Dynavox products can be very difficult for individuals with muscular difficulties (e.g., Parkinson's disease) to accurately press.

9.1   SYMBOLS AND RATE ENHANCEMENT IN TEXT ENTRY

AAC technologies employ a wide range of inputs, including hand gestures, typing, eye movements, and head movements [Glennen and DeCoste, 1996]. These are often made to minimize muscle movement in support of their users' global motor deficits. Slight eye or head movements can be used for both screen-based and screen-free paradigms in which the orientation of the user's gaze or of a head-operated tracking device (such as a joystick or headlamp) is transduced to a cursor position on a screen [Dymond and Potter, 1996]. Typically, items on the screen are selected when the user dwells on them for a time or performs a specific action such as blinking or activating a switch. Screen-based paradigms for AAC are associated with several problems. For flexible communication and mobility, on-screen keyboards and symbol boards often present a large number of targets in a relatively small space, which often requires difficult specific selections, especially for users with severe motor impairments. AAC users have reported, to the author, that screen-based approaches can also interfere with certain social aspects of conversation. In particular, users note that screens often form a barrier to eye contact between conversants, and that, in the case of digital screens, conversation partners will often "read-as-they-go", and guess what a user is trying to say, resulting in editorialization of the user's self-expression.

For these reasons, some technologies employ a screen-free approach to AAC using eye and head movements. Screen-free communication relies on the coding of characters and/or phrases into gestures or sequences of discrete inputs from devices such as switches or joysticks. Among the challenges of designing a screen-free AAC system is how to maximize the accuracy of the user's inputs, while minimizing their effort. The speed with which a character-representing code can be entered is another major design concern. Since screen-free systems tend to require multiple steps to input a single character, these systems tend to be slow.

Figure 9.1: Example of a two-button headswitch mounted on a wheelchair. Image used by permission of the Tetra Society of North America.

Empirical AAC systems for gestural text entry have sought to minimize selection complexity by limiting the number of possible inputs. The H4, EdgeWrite and "Left, Up, Right, Down" Writer systems all rely on codes that are combinations of four discrete inputs [Castellucci and Mackenzie, 2013, Felzer and Nordmann, 2006, Wobbrock et al., 2003], typically target regions placed at the edges or corners of a screen. The MDITIM (Minimal Device Independent Text Input Method) system uses a similar convention, with four inputs dedicated to the coding of characters and one input reserved as a modifier, for example, to achieve capitalization [Isokoski and Raisamo, 2000]. In order to further simplify the input process, the H4 and MDITIM systems, unlike EdgeWrite, have used prefix-free codes to avoid the need for a unique termination event (e.g., a finger-up or blink) to designate the end of each character [Castellucci and Mackenzie, 2013, Isokoski and Raisamo, 2000].

Expert users, with about two and a half hours of experience using the EyeS eye gesture-based communication system, were found to have text communication rates of 6.8 words per minute (wpm), as compared with typical speech rates of 130–200 wpm, and typing rates of 30–40 wpm for competent, unimpaired typists [Arons, 1992, Newell et al., 1998, Porta and Turina, 2008]. Similarly, users with five hours of practice using the MDITIM had an average text entry speed of less than 10 wpm [Isokoski and Raisamo, 2000]. One approach to improving a code's communication rate is to reduce the number of inputs needed to enter each character. The H4 system uses Huffman codes to form a prefix-free code of minimum average length, and has produced an average text entry rate of 20 wpm for users with approximately 6.5 hours of experience [Mackenzie et al.].

Another strategy for improving text entry speed to minimize the number of characters required for entering a word, which can be accomplished by word prediction. Trnka et al. [2008] studied the effect of word prediction on communication rates in an AAC-like onscreen keyboard system. They found that the number of words per minute produced by a user increased with the use of basic word prediction based on a recency-of-use model, as compared with no word prediction. Use of an advanced word prediction algorithm based on statistical natural language modeling proved more effective, yielding an increase in communication rate of 56.8% as compared with no word prediction. The number of options presented in a word prediction list is an important factor in the communication rate of a system. Increasing the length of the list improves the chances that the desired word will be found, but this also increases the visual or auditory scan time to evaluate the list. Mackenzie [2002] suggested that a list size of five is optimal.

9.2 PAYING FOR AAC DEVICES

AAC devices are classified as medical devices in some jurisdictions, and there are various mechanisms that can offset their cost, but their support by governments and large organizations are still, perhaps, in its infancy. Typically, in order to qualify for financial aid toward an AAC device, a formal assessment of ability must be performed by a speech-language pathologist; sometimes, an additional prescription by a physician or family doctor is either sufficient or necessary. Different jurisdictions have different aid mechanisms, but common tools, at least in the United States, are:

Education organizations Certain schools in many jurisdictions are required by law to provide appropriate assistive technology or support to students. In the United States, the Individuals with Disabilities Education Act (IDEA, P.L. 101-476) and its 1997 amendment discuss the inclusion of AT. Generally, services must not cost students or their parents if the Individualized Education Plan team determines that the use of AT is necessary to receive public education.

Medicaid Medicaid is federally funded but implemented and regulated at the state level, and can be managed by an HMO. Medicaid coverage is based on income and/or disability. In general, Medicaid follows Medicare guidelines, but adjustments exist across each state; for example, some states do not require the AAC be 'locked' (i.e., devoted solely to AAC).

Medicare Medicare is the largest funding source in the United States for AAC devices. They are federally funded and administered throughout different *Regional Centers*. Funding is based on necessity, under the category Durable Medical Equipment. Occasionally, their benefits are managed by HMOs, which can complicate things but changes fundamentally little in terms of coverage. Medicare typically covers individuals aged 65 years and over, people with documented disabilities, or family members who have paid Social Security. Medicare allows special exceptions for individuals with amyotrophic lateral sclerosis. Devices such as laptops and mobile devices that are not dedicated AAC devices are not covered.

Private insurance Coverage of AAC devices is dependent on the private insurer, and prior authorization is usually required. Almost all private insurers (e.g., Blue Cross, Kaiser, Premera) require a co-payment of anywhere from 50% to 100% of the device, which is the responsibility of the policy holder. Policies can vary greatly even within a single insurer with regard to coverage of AAC devices.

Tricare Formerly called "CHAMPUS", Tricare is a program for dependents of military service members (both active duty and veteran). The American Congress expanded Tricare support for speech generating devices in 2002.

9.3 DEVICES THAT GENERATE SPEECH

Generating intelligible speech signals for individuals who either have little or no speech function is one of the most fundamental tasks of electronic communication. These systems can be made to synthesize speech given text entered from any of the modalities described in Section 9.1. There are various issues which must be considered and part of any such system, and those are surveyed in the following sections.

9.3.1 USAGE SCENARIO

Consider a dysarthric individual who is traveling into a city by transit to attend an appointment. This might involve purchasing tickets, asking for directions, or indicating their presence or intention to fellow passengers, which must often be done in a noisy and crowded environment. A personal portable communication device in this scenario (either held or attached to a wheelchair) would transform relatively unintelligible speech spoken into a microphone and play the results of that transformation over a set of speakers so that it could be better understood by a listener in that environment. Such a system could facilitate interaction and overcome difficult or failed attempts at communication in daily life. Such an interaction is represented in Figure 9.2.

Figure 9.2: Hypothetical conversation between a speaker with dysarthria (left) and a member of the general population (right). The black box mounted on the wheelchair accepts spoken input and transmits a transformed version of that speech which is more intelligible to the typical hearer.

Similar projects have begun to implement certain aspects of this scenario. In particular, the concept of VOICE BANKING is becoming increasingly popular, where individuals record their speech in the early stages of a degenerative disease (or gender- and aged-matched controls are recruited) to be used in a speech synthesizer [Creer et al., 2009, Jreige et al., 2009, Yamagishi et al., 2012].

Many aspects of daily living are duplicated in order to provide access to individuals with disabilities. The Wheel-Trans service in Toronto,[3] for example, provides accessible transit to persons with physical disabilities who cannot use regular transit and is operated by the city's general transit commission. However, this type of service can be more expensive than the generic systems and it requires booking trips well in advance which can be inconvenient. In many cities, this type of service does not exist at all. Furthermore, such services still require the communication of one's destination or route in which augmentative systems would still be applicable or necessary.

There are augmentative communication devices that employ synthetic text-to-speech in which messages can be written on a specialized keyboard or played back from a repository of pre-recorded messages. However, the type of acoustics produced by such systems often lacks a sufficient degree of individual affectation or natural expression that one might expect in typical human speech [Kain et al., 2007]. The use of *prosody* to convey personal information such as one's emotional state is simply not supported by such systems but is part of a general communicative ability. Transforming one's speech in a way that preserves the natural prosody will therefore also preserve extra-linguistic information, such as emotions, and is therefore a pertinent and crucial response to the limitations of current technology.

9.3.2   FIXED VS. DYNAMIC DISPLAYS

Traditionally, many devices that generate synthetic speech have involved arrays of symbols and other selectable items that are fixed not only in content but also in layout, as exemplified in Figure 9.3. These devices tend to be highly learnable initially [Hochstein et al., 2004] and over time, given so called "muscle memory" and the sheer force of repetition, although there is some evidence that this learnability reaches an asymptote in the long run [Hochstein et al., 2003]. To some extent, these relatively "low tech" devices mimic the fixed nature of older AAC devices such as communication boards. Typically, since the symbols are limited in scope, so are the resulting messages.

---

[3]`https://ttc.ca/WheelTrans/index.jsp`

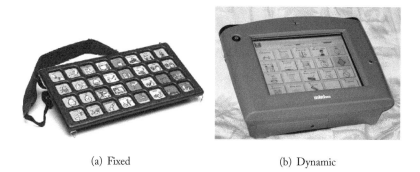

(a) Fixed   (b) Dynamic

Figure 9.3: Static and dynamic display speech-generating devices. (a): Courtesy of National Public Website on Assistive Technology; (b): Courtesy of Poule.

Figure 9.4: Stereotypical arrangement of semantic concepts used in word selection. Courtesy of © 2016 Copyright PrAACtical AAC.

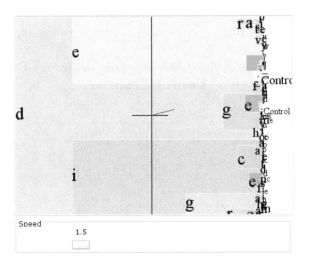

Figure 9.5: Dasher dynamic text entry. The dynamics of this system are based on N-gram models, as described in Section 3.1. Courtesy of David J. C. MacKay.

CHAPTER 10

Supporting Daily Activities Through Speech

Many nations, including most Western nations, are facing major overhauls of their healthcare systems as their populations are tending to be older, which will increase the prevalence of disorders and ailments related to aging. Given the current practice, these nations will not have the capacity to support these older populations or the chronic conditions associated with aging, including dementia. The current healthcare model of removing older adults from their homes and placing them into long-term care facilities is neither financially sustainable (since the proportion of caregivers relative to the growing demand is decreasing) [Bharucha et al., 2009], nor is it desirable. Technologies are being designed to help older adults "age-in-place" by providing different types of support for people who are still able to live at home independently, but require some light assistance.

These "smart home" technologies often use advanced sensors and machine learning to autonomously react to their users, but in each case they tend to be fixed and embedded into the environment, e.g., as cameras in the ceiling. Fixing the location of these technologies carries a tradeoff—installing multiple hardware units at all locations where assistance could be required (e.g., bathroom, kitchen, and bedroom) can be expensive and cumbersome, but installing too few units will limit their utility. Alternatively, integrating personal mobile robots with smart homes can overcome some of these tradeoffs. In fact, assistance provided via a physically embodied robot is often more acceptable than that provided by an embedded system [Klemmer et al., 2006].

One critical component for the successful use of these technological interventions is the usability of the communication interface for the targeted users, in this case older adults with Alzheimer's disease. As in communication between two people, communication between the older adult and the robot may include natural, freeform speech (as opposed to simple spoken keyword interaction) and non-verbal cues (e.g., hand gestures, head pose, eye gaze, facial feature cues), although speech tends to be far more effective [Goodrich and Schultz, 2007, Green et al., 2008]. Previous research indicates that automated communication systems are more effective if they take into account the affective and mental states of the user [Saini et al., 2005]. Indeed, speech appears to be the most powerful mode of communication for an assistive robot to communicate with its users [Lucet, 2012, Tapus and Chetouani, 2010].

10.1   PERSONAL CAREGIVING ROBOTS

Research in smart home systems, assistive robots, and integrated robot/smart home systems for older adults with cognitive impairments has often focused on assistance with activities of daily living (ADL, i.e., reminders to do specific activities according to a schedule or prompts to perform activity steps), cognitive and social stimulation (social interaction with the robot itself or as an interface to communicate with family or friends), and emergency response systems. For smart home systems, Archipel [Serna et al., 2007] recognizes the user's intended plan and provides prompts, e.g., with cooking tasks. Autominder [Pollack, 2006], provides context-appropriate reminders for activity schedules, and the COACH (Cognitive Orthosis for Assisting with aCtivities in the Home) system, being developed by our group, prompts for the task of hand-washing [Mihailidis et al., 2008] and tea-making [Olivier et al., 2009]. Mynatt et al. [2004] have been developing technologies to support aging-in-place such as the Cook's Collage, which uses a series of photos to remind the user what the last step completed was if the user is interrupted during a cooking task. These interventions tend to be embedded in existing environments (e.g., around the sink area).

In assistive robots, researchers have developed socially assistive robots to remind older adults to take medication, eat, use the bathroom, or take a walk [Nejat and Ficocelli, 2008], engage older adults in a memory card game [Nejat and Ficocelli, 2008], and assist during a meal-eating activity [McColl and Nejat, 2013]. Other groups have worked on cognitive and social stimulation activities such as the robotic seal Paro, which is able to respond to touch, sound, sight, and temperature to improve the mood and social engagement of a large sample population of older adults with AD, also decreasing dependency on caregivers [Shibata, 2004]. The Bandit robot is an anthropomorphic upper-torso humanoid robot that encourages exercise and mental activity [Tapus et al., 2008] which was employed in a number of studies where it served as a social companion for older adults with dementia through executing a number of social cues (e.g., pointing, prompting) for playing music, reading books and newspapers, playing a custom game, and motivating mild physical exercise. More recent innovations have examined integrated robot-smart home systems where systems are embedded into existing environments that communicate with mobile assistive robots (e.g., CompanionAble, [Mouad et al., 2010]; Mobiserv Kompai, [Lucet, 2012]; and ROBADOM [Tapus and Chetouani, 2010]). Other work has focused on exercise found that physical embodiment of agents is important. Fasola and Matarić [2013], for example, found from an (albeit relatively small $N = 33$) sample that older adults clearly preferred physically embodied robot coaches over virtual coaches in terms of enjoyableness, helpfulness, social attraction, and related factors.

Many of these projects are targeted toward older adults with cognitive impairment, and not specifically those with significant cognitive impairment. One of these systems, CompanionAble, with a fully autonomous assistive robot, has recently been tested in a simulated home environment for two days each with four older adults with dementia (AD or Pick's disease/frontal lobe dementia) and two with mild cognitive impairment. The system provides assistance with various activities, including appointment reminders for activities input by users or caregivers, video calls, and cognitive exercises. Participants reported an overall acceptance of the system and several upgrades were reported, including a speech recognition system that had to be deactivated by the second day due to poor performance.

One critical component for the successful use of these technological interventions is the usability of the communication interface for the targeted users, in this case older adults with Alzheimer's disease. As in communication between two people, communication between the older adult and the robot may include natural, freeform speech (as opposed to simple spoken keyword interaction) and non-verbal cues (e.g., hand gestures, head pose, eye gaze, facial feature cues), although speech tends to be far more effective [Goodrich and Schultz, 2007, Green et al., 2008]. Previous research indicates that automated communication systems are more effective if they take into account the affective and mental states of the user [Saini et al., 2005]. Indeed, speech appears to be the most powerful mode of communication for an assistive robot to communicate with its users [Lucet, 2012, Tapus and Chetouani, 2010].

In our own work, [Rudzicz et al., 2015], we are building a system that makes steps toward these goals. To help individuals with Alzheimer's disease live at home for longer, we are developing a mobile robot, called ED, intended to assist with activities of daily living through visual monitoring and verbal prompts in cases of difficulty. In a series of experiments, we studied speech-based interactions between ED and each of 10 older adults with Alzheimer's as the latter completed daily tasks in a simulated home environment. That work showed that speech recognition remains a challenge in this setup, especially during household tasks. Across the verbal behaviors that indicate confusion, older adults with AD are very likely to simply ignore the robot, which accounts for over 40% of all such behaviors. However, when they paid attention, older adults were much more likely to have successful conversations (i.e., without instances of confusion) with the robot than with an unfamiliar human conversation partner. This very early work may point the way to more optimized human-robot interaction, which will ideally one day be more sensitive to our personal and emotive needs.

CHAPTER 11

Final Thoughts

Despite the relative speed and ease with which information can be conveyed verbally for most people, for millions of individuals communication remains a severe challenge. Language is fundamental to our society, and it is also the means by which we define ourselves in, and how we engage with, that society. Ensuring that we all have the same basic access, through language, to each other is of the utmost importance to us individually and collectively.

Our species distinguishes itself by its exceptional capacity to understand and to overcome limitations presented by nature, and our technology has given us abilities that we once thought were impossible.

We shape our tools and thereafter our tools shape us.
– Marshall McLuhan [1964]

Bibliography

James H. Abbs, John W. Folkins, and Murali Sivarajan. Motor Impairment following Blockade of the Infraorbital Nerve: Implications for the Use of Anesthetization Techniques in Speech Research. *Journal of Speech and Hearing Research*, 19(1): 19–35, 1976. `http://jslhr.asha.org/cgi/content/abstract/19/1/19`. DOI: 10.1044/jshr.1901.19. 53

S.M. Ahadi-Sarkani. *Bayesian and predictive techniques for speaker adaptation*. Ph.D. thesis, Cambridge University, 1996. 21

Jonathan Allen, M. Sharon Hunnicutt, Dennis H. Klatt, Robert C. Armstrong, and David B. Pisoni. *From text to speech: the MITalk system*. Cambridge University Press, New York, NY, USA, 1987. 30

Norman Alm, John L. Arnott, and Alan F. Newell. Prediction and conversational momentum in an augmentative communication system. *Communications of the ACM*, 35(5): 46–57, 1992. DOI: 10.1145/129875.129878. 13

American Psychiatric Association. Delirium, dementia, and amnestic and other cognitive disorders. In *Diagnostic and Statistical Manual of Mental Disorders, Text Revision (DSM-IV-TR)*, chapter 2. American Psychiatric Association, Arlington, VA, 4th ed., 2000. 41, 42

Julian Appell, Andrew Kertesz, and Michael Fisman. A study of language functioning in {A}lzheimer patients. *Brain and Language*, 17(1): 73–91, 1982. DOI: 10.1016/0093-934x(82)90006-2. 44

Barry Arons. Techniques, perception, and applications of time-compressed speech. In *Proceedings of the 1992 Conference of the American Voice I/O Society*, pages 169–177, 1992. 59

Augmentative Communication Incorporated (ACI). Section 3: Clinical Aspects of AAC Devices, 2007. `http://www.augcominc.com/whatsnew/ncs3.html`. 47

Alan Baddeley and Barbara Wilson. Phonological coding and short-term memory in patients without speech. *Journal of Memory and Language*, 24(4): 490–502, 1985. `http://www.sciencedirect.com/science/article/B6WK4-4D62JF8-G/2/2a2b5ca6878d7761ff586565af5614fb`. DOI: 10.1016/0749-596x(85)90041-5. 54

Alan Baddeley, Susan Gathercole, and Costanza Papagno. The phonological loop as a language learning device. *Psychological Review*, 105(1): 158–173, January 1998. DOI: 10.1037/0033-295x.105.1.158. 54

H.B. Barlow. Unsupervised learning. *Neural Computation*, 1(3): 295–311, 1989. DOI: 10.1162/neco.1989.1.3.295. 53

C. Philip Beaman. Modern cognition in the absence of working memory: Does the working memory account of Neandertal cognition work? *Journal of Human Evolution*, pages 702–706, 2007. DOI: 10.1016/j.jhevol.2006.11.008. 54

Ashok J. Bharucha, Vivek Anand, Jodi Forlizzi, Mary Amanda Dew, Charles F. Reynolds III, Scott Stevens, and Howard Wactlar. Intelligent assistive technology applications to dementia care: Current capabilities, limitations, and future challenges. *American Journal of Geriatric Psychiatry*, 17(2): 88–104, February 2009. DOI: 10.1097/jgp.0b013e318187dde5. 65

Paul Boersma. *Functional Phonology: Formalizing the interactions between articulatory and perceptual drives*. PhD thesis, Universiteit van Amsterdam, September 1998. 27, 28, 29

Paul Boersma. Optimality-theoretic learning in the PRAAT program . In *Proceedings of the Institute of Phonetic Sciences*, volume 23, pages 17–35, 1999. 29

Steven J. Castellucci and I. Scott Mackenzie. Gestural text entry using Huffman codes. In *Proceedings of the International Conference on Multimedia and Human-Computer Interaction*, volume 119, pages 1–8, 2013. 59

Fangxin Chen and Aleksandar Kostov. Optimization of dysarthric speech recognition. In *Proceedings of the 19th Annual International Conference of the IEEE*, volume 4, pages 1436–1439. Engineering in Medicine and Biology society, November 1997. DOI: 10.1109/iembs.1997.756975. 50, 52

Helen Chen and Kenneth N. Stevens. An acoustical study of the fricative /s/ in the speech of individuals with dysarthria. *Journal of Speech, Language, and Hearing Research*, 44: 1300–1314, December 2001. DOI: 10.1044/1092-4388(2001/101). 51

Cecil H. Coker. Speech synthesis with a parametric articulatory model. In *Proceedings of the Speech Symposium*, Kyoto, Japan, 1968. 28

Cecil H. Coker. A model of articulatory dynamics and control. *Proceedings of the IEEE*, 64(4): 452–460, April 1976. DOI: 10.1109/proc.1976.10154. 28

S. M. Creer, P. D. Green, and S. P. Cunningham. Voice banking. *Advances in clinical neuroscience and rehabilitation*, 9(2): 16–17, 2009. 62

David Crystal, *The Cambridge Encyclopedia of Language*, 2nd ed., Cambridge University Presss, Cambridge, UK, 1998. xiv

B. Davis and M. Maclagan. Examining pauses in Alzheimer's discourse. *American journal of Alzheimer's Disease and other dementias*, 24(2): 141–154, 2009. DOI: 10.1177/1533317508328138. 43

J. R. Deller, J. H. L. Hansen, and J. G. Proakis. *Discrete-Time Processing of Speech Signals*. IEEE Press, 2000. DOI: 10.1109/9780470544402. 28

Joseph R Duffy. *Motor Speech Disorders: Substrates, Differential Diagnosis, and Management.* Mosby Inc., 2005. 48

Elizabeth Dymond and Roger Potter. Controlling assistive technology with head movements – a review. *Clinical Rehabilitation*, 10(2): 93–103, 1996. DOI: 10.1177/026921559601000202. 58

Pamela M. Enderby. *Frenchay Dysarthria Assessment.* College Hill Press, 1983. DOI: 10.3109/13682828009112541. 30, 52

Hakan Erdogan, Ruhi Sarikaya, Stanley F. Chen, Yuqing Gao, and Michael Picheny. Using semantic analysis to improve speech recognition performance. *Computer Speech and Language*, 19:321–343, 2005. DOI: 10.1016/j.csl.2004.10.002. 15

J. Fasola and M. J. Matarić. A Socially Assistive Robot Exercise Coach for the Elderly. *Journal of Human-Robot Interaction*, 2(2):3–32, 2013. DOI: 10.5898/jhri.2.2.fasola. 66

Afsaneh Fazly and Graeme Hirst. Testing the efficacy of part-of-speech information in word completion. In *Proceedings of the EACL 2003 Workshop on Language Modeling for Text Entry Methods*, 2003. DOI: 10.3115/1628195.1628197. 15

Torsten Felzer and Rainer Nordmann. Alternative text entry using different input methods. In *Proceedings of the 8th ACM Conference on Computers and Accessibility (ASSETS '06)*, 2006. DOI: 10.1145/1168987.1168991. 59

K. Fraser, F. Rudzicz, and E. Rochon. Using text and acoustic features to diagnose progressive aphasia and its subtypes. In *Proceedings of Interspeech 2013*, pp. 2177–2181, 25–29 August, 2013, Lyon, France.

K. Fraser, J. A. Meltzer, and F. Rudzicz. Linguistic features differentiate Alzheimer's from controls in narrative speech. *Journal of Alzheimer's Disease*, 49(3):407–422, 2015.

Hans-Joachim Freund, Marc Jeannerod, Mark Hallett, and Ramón Leiguarda. *Higher-order motor disorders: From neuroanatomy and neurobiology to clinical neurology.* Oxford University Press, 2005. 52

Nestor Garay-Vitoria and Julio Abascal. Text prediction systems: a survey. *Universal Access in the Information Society*, 4(3):188–203, 2006. DOI: 10.1007/s10209-005-0005-9. 13, 15

Sharon L. Glennen and Denise C. DeCoste. *The Handbook of Augmentative and Alternative Communication.* Singular, San Diego, CA, 1996. 58

R. Goldfarb and M.J.S. Pietro. Support systems: Older adults with neurogenic communication disorders. *Journal of Ambulatory Care Management*, 27(4):356–365, 2004. DOI: 10.1097/00004479-200410000-00008. 46

M. A. Goodrich and A. C. Schultz. Human-robot interaction: A survey. *Foundations and Trends in Human-Computer Interaction*, 1:203–275, 2007. DOI: 10.1561/1100000005. 65, 67

Justine V. Goozee, Bruce E. Murdoch, and Deborah G. Theodoros. Physiological assessment of tongue function in dysarthria following traumatic brain injury. *Logopedics Phoniatrics Vocology*, 26(2):51–65, 2001. DOI: 10.1080/140154301753207421. 49

Vincent L. Gracco. Central and peripheral components in the control of speech movements. In Fredericka Bell-Berti and Lawrence J. Raphael, Eds., *Introducing Speech: Contempory Issues, for Katherine Safford Harris*, chapter 12, pages 417–431. American Institute of Physics press, 1995. DOI: 10.1121/1.417847. 53

S. A. Green, M. Billinghurst, X. Chen, and J. G. Chase. Human-robot collaboration: A literature review and augmented reality approach in design. *International Journal Advanced Robotic Systems*, 5:1–18, 2008. DOI: 10.5772/5664. 65, 67

Frank H. Guenther and Joseph S. Perkell. A neural model of speech production and its application to studies of the role of auditory feedback in speech. In Ben Maassen, Raymond Kent, Herman Peters, Pascal Van Lieshout, and Wouter Hulstijn, Eds., *Speech Motor Control in Normal and Disordered Speech*, chapter 4, pages 29–49. Oxford University Press, Oxford, 2004. 54

Curry Guinn and Anthony Habash. Technical Report FS-12-01, Association for the Advancement of Artificial Intelligence, 2012. 43

Vicki L. Hammen, Kathryn M. Yorkston, and Fred D. Minifie. Effects of Temporal Alterations on Speech Intelligibility in Parkinsonian Dysarthria. *Journal of Speech and Hearing Research*, 37:244–253, 1994. DOI: 10.1044/jshr.3702.244. 49, 52

Mark Hasegawa-Johnson, Jon Gunderson, Adrienne Perlman, and Thomas Huang. HMM-based and SVM-based recognition of the speech of talkers with spastic dysarthria. In *Proceedings of the International Conference on Acoustics, Speech, and Signal Processing (ICASSP 2006)*, volume 3, pages 1060–1063, May 2006a. DOI: 10.1109/icassp.2006.1660840. 52

Mark Hasegawa-Johnson, Jon Gunderson, Adrienne Perlman, and Thomas S. Huang. Audiovisual phonologic-feature-based recognition of dysarthric speech. abstract, 2006b. 48

Wolfgang J. Hess. Pitch and voicing determination of speech with an extension toward music signal. In Jacob Benesty, M. Mohan Sondhi, and Yiteng Huang, Eds., *Speech Processing*. Springer, 2008. 30

Dave D. Hochstein, Mark a. McDaniel, Sandra Nettleton, and Katherine Hannah Neufeld. The fruitfulness of a nomothetic approach to investigating AAC: Comparing two speech encoding schemes across cerebral palsied and nondisabled children. *American Journal of Speech-Language Pathology*, 12(1):110–120, 2003. DOI: 10.1044/1058-0360(2003/057). 62

Dave D. Hochstein, Mark A. McDaniel, and Sandra Nettleton. Recognition of Vocabulary in Children and Adolescents with Cerebral Palsy: A Comparison of Two Speech Coding Schemes. *Augmentative and Alternative Communication*, 20(2):45–62, 2004. http://informahealthcare.com/doi/abs/10.1080/07434610410001699708. DOI: 10.1080/07434610410001699708. 62

T Hopper. Indirect interventions to facilitate communication in Alzheimer's disease. *Seminars in Speech and Language*, 22(4):305–315, 2001. DOI: 10.1055/s-2001-17428. 46

John-Paul Hosom, Alexander B. Kain, Taniya Mishra, Jan P. H. van Santen, Melanie Fried-Oken, and Janice Staehely. Intelligibility of modifications to dysarthric speech. In *Proceedings of the IEEE International Conference on Acoustics, Speech, and Signal Processing (ICASSP '03)*, volume 1, pages 924–927, April 2003. DOI: 10.1109/icassp.2003.1198933. 30, 58

Xuedong Huang, Alex Acero, and Hsiao-Wuen Hon. *Spoken Language Processing: A Guide to Theory, Algorithm and System Development*. Prentice Hall PTR, April 2001. http://www.amazon.co.uk/exec/obidos/ASIN/0130226165/citeulike-21. 20, 21, 22, 27, 28, 29

Marco F. Huber, Tim Bailey, Hugh Durrant-Whyte, and Uwe D. Hanebeck. On entropy approximation for Gaussian mixture random vectors. In *Proceedings of the 2008 IEEE International Conference on In Multisensor Fusion and Integration for Intelligent Systems*, pages 181–188, Seoul, South Korea, 2008. DOI: 10.1109/mfi.2008.4648062. 12

Katherine C. Hustad. Estimating the intelligibility of speakers with dysarthria. *Folia Phoniatrica et Logopaedica*, 58(3):217–228, 2006. DOI: 10.1159/000091735. 30

Khalil Iskarous, Louis M. Goldstein, D.H. Whalen, Mark K. Tiede, and Philip E. Rubin. CASY: The Haskins Configurable Articulatory Synthesizer. In *Proceedings of the 15th International Congress of Phonetic Sciences*, pages 185–188, Barcelona, Spain, August 2003. 29

Poika Isokoski and Roope Raisamo. Device independent text input: A rationale and an example. In *Proceedings of the ACM Working Conference on Advanced Visual Interfaces (AVI '00)*, pages 76–83, 2000. DOI: 10.1145/345513.345262. 59

Camil Jreige, Rupal Patel, and H. Timothy Bunnell. Vocalid: Personalizing text-to-speech synthesis for individuals with severe speech impairment. In *Proceedings of the 11th International ACM SIGACCESS Conference on Computers and Accessibility*, Assets '09, pages 259–260, New York, NY, USA, 2009. ACM. http://doi.acm.org/10.1145/1639642.1639704. DOI: 10.1145/1639642.1639704. 62

Daniel Jurafsky and James H. Martin. *Speech and Language Processing: An introduction to natural language processing, computational linguistics, and speech recognition*. Prentice Hall, 2nd ed., 2009. 15, 19, 20, 40

Alexander B. Kain, John-Paul Hosom, Xiaochuan Niu, Jan P.H. van Santen, Melanie Fried-Oken, and Janice Staehely. Improving the intelligibility of dysarthric speech. *Speech Communication*, 49(9):743–759, September 2007. DOI: 10.1016/j.specom.2007.05.001. 30, 31, 62

Ray D. Kent. Research on speech motor control and its disorders: a review and prospective. *Journal of Communication Disorders*, 33(5):391–428, 2000. DOI: 10.1016/s0021-9924(00)00023-x. 50, 52

Ray D. Kent and Kristin Rosen. Motor control perspectives on motor speech disorders. In Ben Maassen, Raymond Kent, Herman Peters, Pascal Van Lieshout, and Wouter Hulstijn, Eds., *Speech Motor Control in Normal and Disordered Speech*, chapter 12, pages 285–311. Oxford University Press, Oxford, 2004. 47, 49, 52

Ray D. Kent, Gary Weismer, Jane F. Kent, and John C. Rosenbek. Toward phonetic intelligibility testing in dysarthria. *Journal of Speech and Hearing Disorders*, 54:482–499, 1989. DOI: 10.1044/jshd.5404.482. 52

Yusuke Kida and Tatsuya Kawahara. Voice activity detection based on optimally weighted combination of multiple features. In *Proceedings of INTERSPEECH-2005*, pages 2621–2624, 2005. 30

Heejin Kim, Mark Hasegawa-Johnson, and Adrienne Perlman. Acoustic cues to lexical stress in spastic dysarthria. In *Proceedings of Speech Prosody 2010*, 2010a. 51

Heejin Kim, Panying Rong, Torrey M. Loucks, and Mark Hasegawa-Johnson. Kinematic analysis of tongue movement control in spastic dysarthria. In *Proceedings of Interspeech 2010*, 2010b. 50

H. S. Kirshner. Primary progressive aphasia and Alzheimer's disease: brief history, recent evidence. *Current Neurology and Neuroscience Reports*, 12:709–714, 2012. DOI: 10.1007/s11910-012-0307-2. 44

Dennis H. Klatt. Software for a cascade/parallel formant synthesizer. *Journal of the Acoustical Society of America*, 67(3):971–995, 1980. DOI: 10.1121/1.383940. 29

S. Klemmer, B. Hartmann, and L. Takayama. How bodies matter: five themes for interaction design. In *Proceedings of the conference on Designing Interactive systems*, pages 140–149, 2006. DOI: 10.1145/1142405.1142429. 65

Aida C. G. Verdugo Lazo and Pushpa N. Rathie. On the entropy of continuous probability distributions. *IEEE Transactions on Information Theory*, 23(1):120–122, January 1978. DOI: 10.1109/tit.1978.1055832. 12

Jianhua Li and Graeme Hirst. Semantic knowledge in word completion. In *Assets '05: Proceedings of the 7th international ACM SIGACCESS conference on Computers and accessibility*, pages 121–128, New York, NY, USA, 2005. ACM Press. DOI: 10.1145/1090785.1090809. 15

William John Little. On the influence of abnormal parturition, difficult labour, premature birth, and asphyxia neonatorum on the mental and physical condition of the child, especially in relation to deformities. *Transactions of the Obstetrical Society of London*, 3:293–344, 1861. DOI: 10.1001/archneur.1969.00480080118015. 47

Anders Löfqvist, Nancy S. McGarr, and Kiyoshi Honda. Laryngeal muscles and articulatory control. *The Journal of the Acoustical Society of America*, 76(3):951–954, 1984. http://link.a ip.org/link/?JAS/76/951/1. DOI: 10.1121/1.391278. 37

Eric Lucet. Social Mobiserv Kompai Robot to Assist People. In *euRobotics workshop on Robots in Healthcare and Welfare*, 2012. 65, 66, 67

I. Scott Mackenzie. KSPC as a characteristic of text entry techniques. In *Proceedings of the 4th ACM International Conference on Human-Computer Interaction with Mobile Devices and Services (MobileHCI '02)*, pages 195–210, 2002. 60

I. Scott Mackenzie, R. William Soukoreff, and Joanne Helga. 1 Thumb, 4 Buttons, 20 Words Per Minute: Design and Evaluation of H4-Writer. In *Proceedings of the 24th Annual ACM Conference on User Interface Software and Technology (UIST '11)*, pages 471–480. DOI: 10.1145/2047196.2047258. 59

Shinji Maeda. Compensatory articulation during speech: evidence from the analysis and synthesis of vocal-tract shapes using an articulatory model. In W.J. Hardcastle and A. Marchal, Eds., *Speech Production and Speech Modelling*, pages 131–149. Kluwer, 1990. DOI: 10.1007/978-94-009-2037-8. 29

Johannes Matiasek, Marco Baroni, and Harald Trost. FASTY : A Multi-lingual Approach to Text Prediction. In *ICCHP '02: Proceedings of the 8th International Conference on Computers Helping People with Special Needs*, pages 243–250, London, UK, 2002. Springer-Verlag. DOI: 10.1007/3-540-45491-8_51. 13

Derek McColl and Goldie Nejat. Meal-time with a socially assistive robot and older adults at a long-term care facility. *Journal of Human-Robot Interaction*, 2(1):152–171, 2013. DOI: 10.5898/jhri.2.1.mccoll. 66

Monica A. McHenry and Julie M. Liss. The impact of stimulated vocal loudness on nasalance in dysarthria. *Journal of Medical Speech-Language Pathology*, 14(3):197–205, September 2006. 49

Sean McLennan. Klatt Synthesizer in Simulink. Technical report, Indiana University, April 2000. 29

Marshall McLuhan. *Understanding Media: The Extensions of Man*. McGraw Hill, New York NY, 1964. 69

Xavier Menendez-Pidal, James B. Polikoff, Shirley M. Peters, Jennie E. Leonzjo, and H.T. Bunnell. The Nemours Database of Dysarthric Speech. In *Proceedings of the Fourth International Conference on Spoken Language Processing*, Philadelphia PA, USA, October 1996. DOI: 10.1109/icslp.1996.608020. 52

Kinfe T. Mengistu and Frank Rudzicz. Comparing humans and automatic speech recognition systems in recognizing dysarthric speech. In *Proceedings of the Canadian Conference on Artificial Intelligence*, St. John's Canada, May 2011. DOI: 10.1007/978-3-642-21043-3_36. 52

P. Mermelstein. Articulatory model for the study of speech production. *Journal of the Acoustical Society of America*, 53(4):1070–1082, 1973. DOI: 10.1121/1.1913427. 29

Alex Mihailidis, Jennifer N Boger, Tammy Craig, and Jesse Hoey. The COACH prompting system to assist older adults with dementia through handwashing: An efficacy study. *BMC Geriatrics*, 8(28), 2008. DOI: 10.1186/1471-2318-8-28. 66

Keith L. Moore and Arthur F. Dalley. *Clinically Oriented Anatomy*, 5th ed., Lippincott, Williams and Wilkins, 2005. 37

Hiroki Mori, Yasunori Kobayashi, Hideki Kasuya, Noriko Kobayashi, and Hajime Hirose. Evaluation of fundamental frequency (f0) characteristics of speech in dysarthrias: A comparative study. *Acoustical Science and Technology*, 26(6):540–543, 2005. DOI: 10.1250/ast.26.540. 51

Mehdi Mouad, Lounis Adouane, Pierre Schmitt, Djamel Khadraoui, Benjamin Gâteau, and Philippe Martinet. Multi-agents based system to coordinate mobile teamworking robots. In *Proceedings of the 4th Companion Robotics Institute*, Brussels, 2010. 66

Eric Moulines and Francis Charpentier. Pitch-synchronous waveform processing techniques for text-to-speech synthesis using diphones. *Speech Commun.*, 9:453–467, December 1990. http://portal.acm.org/citation.cfm?id=116058.116064. DOI: 10.1016/0167-6393(90)90021-z. 27

Elizabeth D. Mynatt, Anne-Sophie Melenhorst, Arthur D. Fisk, and Wendy A. Rogers. Aware technologies for aging in place: Understanding user needs and attitudes. *IEEE Pervasive Computing*, 3:36–41, 2004. DOI: 10.1109/mprv.2004.1316816. 66

Goldie Nejat and Maurizio Ficocelli. Can i be of assistance? the intelligence behind an assistive robot. In *IEEE International Conference on Robotics and Automation*, pages 3564–3569, 2008. DOI: 10.1109/robot.2008.4543756. 66

Alan Newell, Stefan Langer, and Marianne Hickey. The role of natural language processing in alternative and augmentative communication. *Natural Language Engineering*, 4(1):1–16, 1998. DOI: 10.1017/s135132499800182x. 59

Masaki Nishio and Seiji Niimi. Speaking rate and its components in dysarthric speakers. *Clinical Linguistics & Phonetics*, 15(4):309–317, June 2001. DOI: 10.1080/02699200010024456. 48

Patrick Olivier, Andrew Monk, Guangyou Xu, and Jesse Hoey. Ambient kitchen: Designing situation services using a high fidelity prototyping environment. In *Proceedings of the ACM 2nd International Conference on Pervasive Technologies Related to Assistive Environments*, Corfu Greece, 2009. DOI: 10.1145/1579114.1579161. 66

J. B. Orange, Rosemary B. Lubinsky, and D. Jeffery Higginbotham. Conversational repair by individuals with dementia of the alzheimer's type. *Journal of Speech and Hearing Research*, 39: 881–895, August 1996. DOI: 10.1044/jshr.3904.881. 44

Douglas O'Shaughnessy. *Speech Communications – Human and Machine*. IEEE Press, New York, NY, USA, 2000. 26, 30, 36, 39, 50, 51, 52

Douglas O'Shaughnessy. Formant estimation and tracking. In Jacob Benesty, M. Mohan Sondhi, and Yiteng Huang, Eds., *Speech Processing*. Springer, 2008. 30

Yoshiaki Ozawa, Osamu Shiromoto, Fumiko Ishizaki, and Toshiko Watamori. Symptomatic differences in decreased alternating motion rates between individuals with spastic and with ataxic dysarthria: An acoustic analysis. *International Journal of Phoniatrics, Speech Therapy and Communication Pathology*, 53(2), 2001. DOI: 10.1159/000052656. 48

S. V. Pakhomov, G. E. Smith, D. Chacon, Y. Feliciano, N. Graff-Radford, R. Caselli, and D. S. Knopman. Computerized analysis of speech and language to identify psycholinguistic correlates of frontotemporal lobar degeneration. *Cognitive and Behavioral Neurology*, 23:165–177, 2010. DOI: 10.1097/wnn.0b013e3181c5dde3. 43

Rebecca Palmer, Pam Enderby, and Mark Hawley. Addressing the needs of speakers with long-standing dysarthria: computerized and traditional therapy compared. *International Journal of Language & Communication Disorders*, 42:61–79, 2007. DOI: 10.1080/13682820601173296. 52

Rupal Patel. Control of prosodic parameters by an individual with severe dysarthria. Technical report, University of Toronto, December 1998. http://vismod.media.mit.edu/pub/masters_paper.doc. 49

Rupal Patel. Phonatory control in adults with cerebral palsy and severe dysarthria. *AAC Augmentative and Alternative Communication*, 18:2–10, 2002a. DOI: 10.1080/aac.18.1.2.10. 51

Rupal Patel. Prosodic control in severe dysarthria: Preserved ability to mark the question-statement contrast. *Journal of speech, language, and hearing research*, 45(5):858–870, 2002b. DOI: 10.1044/1092-4388(2002/069). 51

M. E. Pollack. Autominder: A case study of assistive technology for elders with cognitive impairment. *Generations*, 30:67–69, 2006. 66

Prasad D. Polur and Gerald E. Miller. Investigation of an HMM/ANN hybrid structure in pattern recognition application using cepstral analysis of dysarthric (distorted) speech signals. *Medical Engineering and Physics*, 28(8):741–748, October 2006. DOI: 10.1016/j.medengphy.2005.11.002. 51

Marco Porta and Matteo Turina. Eye-S: a Full-Screen Input Modality for Pure Eye-Based Communication. In *Proceedings of the ACM 2008 Symposium on Eye Tracking Research and Applications (ETRA '08)*, pages 27–34, 2008. DOI: 10.1145/1344471.1344477. 59

William Campbell Posey. Some ocular phases of Litte's disease (congenital spastic rigidity of the limbs. *Journal of the American Medical Association*, 80(2):80–82, 1923. DOI: 10.1001/jama.1923.02640290010003. 47

William H. Press, Saul A. Teukolsky, William T. Vetterling, and Brian P. Flannery. *Numerical Recipes in C: the art of scientific computing*. Cambridge University Press, 2nd ed., 1992. 30

Thomas F. Quatieri. *Discrete-time Speech Signal Processing: Principles and Practice*. Prentice Hall, 2002. 19, 27, 28

Parimala Raghavendra, Elisabet Rosengren, and Sheri Hunnicutt. An investigation of different degrees of dysarthric speech as input to speaker-adaptive and speaker-dependent recognition systems. *Augmentative and Alternative Communication (AAC)*, 17(4):265–275, December 2001. DOI: 10.1080/aac.17.4.265.275. 49

Alexander M. Rapp and Barbara Wild. Nonliteral language in Alzheimer dementia: a review. *Journal of the International Neuropsychological Society : JINS*, 17(2):207–218, 2011. DOI: 10.1017/s1355617710001682. 44

J. Reilly, J. Troche, and M. Grossman. Language processing in dementia. In A. E. Budson and N. W. Kowall, Eds., *The Handbook of Alzheimer's Disease and Other Dementias*. Wiley-Blackwell, 2011. DOI: 10.1002/9781444344110. 43

Korin Richmond, Simon King, and Paul Taylor. Modelling the uncertainty in recovering articulation from acoustics. *Computer Speech and Language*, 17:153–172, 2003. DOI: 10.1016/s0885-2308(03)00005-6. 53

Brian Roark, Margaret Mitchell, John-Paul Hosom, Kristy Hollingshead, and Jeffery Kaye. Spoken language derived measures for detecting mild cognitive impairment. *IEEE Transactions on Audio, Speech, and Language Processing*, 19(7):2081–2090, 2011. DOI: 10.1109/tasl.2011.2112351. 43

Keith M. Robinson, Murray Grossman, Tammy White-Devine, and Mark D'Esposito. Category-specific difficulty naming with verbs in {A}lzheimer's disease. *Neurology*, 47(1):178–182, 1996. DOI: 10.1212/wnl.47.1.178. 44

Kristin Rosen and Sasha Yampolsky. Automatic speech recognition and a review of its functioning with dysarthric speech. *Augmentative & Alternative Communication*, 16 (1):48–60, Jan 2000. http://dx.doi.org/10.1080/07434610012331278904. DOI: 10.1080/07434610012331278904. 47, 49

Philip Rubin, Thomas Baer, and Paul Mermelstein. An articulatory synthesizer for perceptual research. *The Journal of the Acoustical Society of America*, 70(2):321–328, 1981. http://link.aip.org/link/?JAS/70/321/1. DOI: 10.1121/1.386780. 29

F. Rudzicz, G. Hirst, and P. Van Lieshout. Vocal tract representation in the recognition of cerebral palsied speech. In *Journal of Speech, Language, and Hearing Research*, 55(4):1190–1207, August, 2012.

F. Rudzicz, R. Wang, M. Begum, and A. Mihailidis. Speech interaction with personal assistive robots supporting aging-at-home for individuals with Alzheimer's disease. *ACM Transactions on Accessible Computing*, 7(2), 2015. DOI: 10.1145/2744206. 67

Frank Rudzicz. Articulatory knowledge in the recognition of dysarthric speech. In *IEEE Transactions on Audio, Speech, and Language Processing*, 19(4), May, 2011, pages 947–960.

Frank Rudzicz. Acoustic transformations to improve the intelligibility of dysarthric speech. In *Proceedings of the Second Workshop on Speech and Language Processing for Assistive Technologies (SLPAT2011)* at the ACL Conference on Empirical Methods in Natural Language Processing (EMNLP2011), 30 July, 2011, Edinburgh, Scotland.

Frank Rudzicz. Adjusting dysarthric speech signals to be more intelligible. *Computer Speech and Language*, 27(6):1163–1177, 2013. DOI: 10.1016/j.csl.2012.11.001. 58

Privender Saini, Boris de Ruyter, Panos Markopoulos, and Albert van Breemen. Benefits of social intelligence in home dialogue systems. In *Proceedings of INTERACT 2005*, pages 510–521, 2005. DOI: 10.1007/11555261_42. 46, 65, 67

Emanuel A. Schegloff, Gail Jefferson, and Harvey Sacks. The preference for self-correction in the organization of repair in conversation. *1977*, 53(2):361–382, 1977. DOI: 10.2307/413107. 44

Carl R. Schneiderman and Robert E. Potter. *Speech-language pathology : a simplified guide to structures, functions, and clinical implications*. Academic Press, San Diego, CA, 2002. 38

Juergen Schroeter. Basic principles of speech synthesis. In Jacob Benesty, M. Mohan Sondhi, and Yiteng Huang, Eds., *Speech Processing*. Springer, 2008. 27, 30

J. Anthony Seikel, Douglas W. King, and David G. Drumright, Eds., *Anatomy & Physiology: for Speech, Language, and Hearing*. Thomson Delmar Learning, 3rd ed., 2005. 53

A. Serna, H. Pigot, and V. Rialle. Modeling the progression of alzheimer's disease for cognitive assistance in smart homes. *User Modelling and User-Adapted Interaction*, 17:415–438, 2007. DOI: 10.1007/s11257-007-9032-y. 66

Claude E. Shannon. *A Mathematical Theory of Communication*. University of Illinois Press, Urbana, IL, 1949. DOI: 10.1002/j.1538-7305.1948.tb01338.x. 9

Takanori Shibata. An overview of human interactive robots for psychological enrichment. *Proceedings of the IEEE*, 92(11):1749–1758, 2004. DOI: 10.1109/jproc.2004.835383. 66

Jeff A. Small, Gloria Gutman, Saskia Makela, and Beth Hillhouse. Effectiveness of communication strategies used by caregivers of persons with alzheimer's disease during activities of daily living. *Journal of Speech, Language, and Hearing Research*, 46(2):353–367, 2003. DOI: 10.1044/1092-4388(2003/028). 46

M. Snover, B. Dorr, and R. Schwartz. A lexically-driven algorithm for disfluency detection. In '*Proceedings of HLT-NAACL 2004: Short Papers*, pages 157–160, 2004. DOI: 10.3115/1613984.1614024. 43

Nancy P. Solomon, Donald A. Robin, and Erich S. Luschei. Strength, Endurance, and Stability of the Tongue and Hand in Parkinson Disease. *Journal of Speech, Language, and Hearing Research*, 43:256–267, 2000. DOI: 10.1044/jslhr.4301.256. 49

Richard A. Spears and Anders Holtz. *Spinal Cord Injury*. Oxford University Press, Oxford UK, 2010. 57

Murray F. Spiegel, Mary Jo Altom, Marian J. Macchi, and Karen L. Wallace. Comprehensive assessment of the telephone intelligibility of synthesized and natural speech. *Speech Communication*, 9(4):279 – 291, 1990. http://www.sciencedirect.com/science/article/B6V1C-48V21K0-FN/2/724c382576c44981bcc8e8bba88626bd. DOI: 10.1016/0167-6393(90)90004-s. 30

Kenneth N. Stevens. *Acoustic Phonetics*. MIT Press, Cambridge, Massachussetts, 1998. 18, 19, 39

S. S. Stevens, J. Volkman, and E. B. Newman. A scale for the measurement of the psychological magnitude pitch. *Journal of the Acoustical Society of America*, 8(3):185–190, January 1937. DOI: 10.1121/1.1915893. 20

Johan Sundberg. The acoustics of the singing voice. *Scientific American*, 234:82–91, 1977. DOI: 10.1038/scientificamerican0377-82. 37

Andrew Swiffin, John Arnott, J. Adrian Pickering, and Alan Newell. Adaptive and predictive techniques in a communication prosthesis. *Augmentative & Alternative Communication*, 3(4): 181–191, December 1987. DOI: 10.1080/07434618712331274499. 13

Adriana Tapus and Mohamed Chetouani. ROBADOM: the impact of a domestic robot on the psychological and cognitive state of the elderly with mild cognitive impairment. In *Proceedings of the International Symposium on Quality of Life Technology Intelligent Systems for Better Living*, Las Vegas USA, June 2010. 65, 66, 67

Adriana Tapus, Juan Fasola, and Maja J Matarić. Cognitive assistance and physical therapy for dementia patients. In *ICRA Workshop on Social Interaction with Intelligent Indoor Robots*, 2008. 66

Nancy Thomas-Stonell, Ava-Lee Kotler, Herbert A. Leeper, and Philip C. Doyle. Computerized speech recognition: influence of intelligibility and perceptual consistency on recognition accuracy. *Augmentative & Alternative Communication*, 14(1):51–56, March 1998. DOI: 10.1080/07434619812331278196. 52

E. C. Thompson, B. E. Murdoch, and P. D. Stokes. Lip function in subjects with upper motor neuron type dysarthria following cerebrovascular accidents. *European Journal of Disorders of Communication*, 30:451–466, 1995. DOI: 10.3109/13682829509087244. 49

Nuttakorn Thubthong, Prakasith Kayasith, Sriwimon Manochiopinig, Wisit Leelasiriwong, and Onwadee Rukkharangsarit. Articulation analysis of Thai cerebral palsy children with dysarthric speech. In *Proceedings of the 6th Symposium on Natural Language Processing*, 2005. 50

Ingo R. Titze. *Principles of Voice Production*. Prentice-Hall, Englewood Cliffs, NJ, 1994. DOI: 10.1121/1.424266. 37

Tomoki Toda, Alan W. Black, and Keiichi Tokuda. Statistical mapping between articulatory movements and acoustic spectrum using a Gaussian mixture model. *Speech Communication*, 50 (3):215–227, March 2008. DOI: 10.1016/j.specom.2007.09.001. 53

Keith Trnka, John McCaw, Debra Yarrington, Kathleen F. McCoy, and Christopher Pennington. Word Prediction and Communication Rate in AAC. In *Proceedings of the Fourth IASTED Conference on Telehealth and Assistive Technologies (Telehealth/AT 2008)*, 2008. 59

Ying-Chiao Tsao, Gary Weismer, and Kamran Iqbal. The effect of intertalker speech rate variation on acoustic vowel space. *The Journal of the Acoustical Society of America*, 119(2):1074–1082, February 2006. DOI: 10.1121/1.2149774. 49

Alan Turing. Computing machinery and intelligence. *Mind*, 59:433–460, 1950. DOI: 10.1093/mind/lix.236.433. 13

T. Umapathi, N. Venketasubramanian, K. J. Leck, C.B. Tan, W.L. Lee, and H. Tjia. Tongue deviation in acute ischaemic stroke: a study of supranuclear twelfth cranial nerve palsy in 300 stroke patients. *Cerebrovascular Diseases*, 10:462–465, 2000. DOI: 10.1159/000016108. 49

Ron Walls, John J. Ratey, and Robert I. Simon. *Rosen's Emergency Medicine: Expert Consult (Premium ed.)*. Mosby, St. Louis Missouri, 2009. 57

Caroline M. Watson. An analysis of trouble and repair in the natural conversations of people with dementia of the Alzheimer's type. *Aphasiology*, 13(3):195 – 218, 1999. DOI: 10.1080/026870399402181. 44

F. G. Weijnen, J. B. M. Kuks, A. van der Bilt, H. W. van der Glas, M. W. Wassenberg, and F. Bosman. Tongue force in patients with myasthenia gravis. *Acta Neurologica Scandinavica*, 102(5):303–308, 2000. DOI: 10.1034/j.1600-0404.2000.102005303.x. 49

Jacob O. Wobbrock, Brad A. Myers, and John A. Kembel. EdgeWrite: A Stylus-Based Text Entry Method Designed for High Accuracy and Stability of Motion. In *Proceedings of the 16th Annual ACM Conference on User Interface Software and Technology (UIST '03)*, pages 61–70, 2003. DOI: 10.1145/964696.964703. 59

Junichi Yamagishi, Christophe Veaux, Simon King, and Steve Renals. Speech synthesis technologies for individuals with vocal disabilities: Voice banking and reconstruction. *Acoustical Science and Technology*, 33(1):1–5, 2012. http://www.jstage.jst.go.jp/browse/ast/33/1/_contents. DOI: 10.1250/ast.33.1. 62

Kathryn M. Yorkston and David R. Beukelman. *Assessment of Intelligibility of Dysarthric Speech*. C.C. Publications Inc., Tigard, Oregon, 1981. 30

Heiga Zen, Keiichi Tokuda, and Alan W. Black. Statistical parametric speech synthesis. *Speech Communication*, 51:1039–1064, 2009. http://dx.doi.org/10.1016/j.specom.2009.04.004. DOI: 10.1016/j.specom.2009.04.004. 24

Heiga Zen, Andrew Senior, and Mike Schuster. Statistical parametric speech synthesis using deep neural networks. *International Conference on Acoustics, Speech and Signal Processing (ICASSP 2013)*, pages 7962–7966, 2013. DOI: 10.1109/icassp.2010.5495691. 24

Wolfram Ziegler and Ben Maassen. *The role of the syllable in disorders of spoken language production*, chapter 16, pages 415–447. Speech Motor Control in Normal and Disordered Speech. Oxford University Press, Oxford, 2004. 51

Author's Biography

FRANK RUDZICZ

Frank Rudzicz is an international expert on speech technology for individuals with speech disorders. He is the president of the international joint ACL/ISCA special interest group on Speech and Language Processing for Assistive Technology and Young Investigator of the Alzheimer's Society. His work involves machine-learning, human-computer interaction, speech-language pathology, rehabilitation engineering, signal processing, and linguistics. Contributions include the TORGO database of disordered speech (the first speech recognition system for people with speech disorders based on their physical speech articulation) and subsequent communication aid software that modifies the acoustics of hard-to-understand speech to make it more understandable to the typical listener. He is the director of SPOClab at the Toronto Rehabilitation Institute and University of Toronto, and a co-founder of WinterLight Labs Inc.

Printed in the United States
by Baker & Taylor Publisher Services